THE

DNA
RESTART

THE

DNA
RESTART

UNLOCK YOUR PERSONAL GENETIC CODE
TO EAT FOR YOUR GENES,
Lose Weight, and Reverse Aging

SHARON MOALEM, MD, PhD

NEW YORK TIMES BESTSELLING AUTHOR OF *SURVIVAL OF THE SICKEST*

Foreword by Chef Nobu Matsuhisa

RODALE.

RODALE *wellness*

Live happy. Be healthy. Get inspired.

Sign up today to get exclusive access to our authors, exclusive bonuses, and the most authoritative, useful, and cutting-edge information on health, wellness, fitness, and living your life to the fullest.

Visit us online at RodaleWellness.com
Join us at RodaleWellness.com/Join

Library of Congress Cataloging-in-Publication Data is on file with the publisher
ISBN 978–1–62336–668–1   trade hardcover
Distributed to the trade by Macmillan
2 4 6 8 10 9 7 5 3 1   hardcover

We inspire health, healing, happiness, and love in the world. Starting with you.

To all DNA Restarters past and present,
for believing real change is possible

CONTENTS

PART I. THE DNA RESTART 1ST PILLAR:
Eat for Your Genes • 1

PART II. THE DNA RESTART 2ND PILLAR:
Reverse Aging • 53

PART III. THE DNA RESTART 3RD PILLAR:
Eat Umami • 127

PART IV. THE DNA RESTART 4TH PILLAR:
Drink Oolong Tea • 163

PART V. THE DNA RESTART 5TH PILLAR:
Slow Living • 185

PART VI. THE DNA RESTART ROAD MAP TO
OPTIMAL HEALTH AND LONGEVITY • 201

FOREWORD

I'll never forget the first time I met Sharon at Nobu Tokyo. I invited him to my restaurant because I wanted to show him firsthand how I use umami, the fifth human taste, in my original dishes. It was really exciting for me to meet a physician and scientist who was so obviously passionate about human health but, like me, also deeply loves food.

My conversation with Sharon that night focused on umami. We've all tasted umami. We just didn't know it at the time. It's found in very common kitchen staples like tomatoes and Parmesan cheese. Now many of the world's top chefs are trying to understand and use umami in their kitchens. Umami helps chefs be more creative, and the dishes they come up with taste better than ever because of it. I've always said to young chefs in my restaurants that delicious can be good for you, allowing food to be low in calories but high in satisfaction. Sharon and I are part of this new culinary movement that's seeking to marry delicious food with improving human health.

As Sharon so clearly demonstrates in his book, if we learn how to use umami correctly, it can be a powerful tool to get us to feel and stay full longer, which can get us all to an ideal weight faster. This idea will spark a dietary revolution.

But the DNA Restart isn't just about delicious food—it's about something much more important. And that's getting every one of us to start eating for our genes. What excites me about Sharon's book is that instead of focusing on food exclusion, it's about bringing us back to a pleasurable eating and living experience using our DNA as the guide. This book is filled with genetic self-tests and the information you need to know concerning which foods and drinks you should be consuming for your genes.

The DNA Restart is an invitation for you to start a journey of genetic discovery that will change your current mind-set about health and longevity.

Cheers for your enjoyment and health!

Nobu Matsuhisa
Los Angeles, 2016

INTRODUCTION

Claire[1] was in her late thirties, working in publishing, and already a well-seasoned veteran dieter when I first met her. While some of her friends and even family members found some success with certain diets, Claire never seemed to be able to find that one diet that worked for her. She told me that her biggest challenge was striking a good balance among her carbohydrate, protein, and fat intake.

To rein in a creeping increase in her dress size and the accompanying 15 pounds she had gained in the last few years, Claire had diligently tried to severely limit her carb intake. What was most disturbing to her, though, was that the more she limited her carbs, the worse she'd feel. Without eating virtually any carbohydrates at all during the day, she found it nearly impossible to concentrate on getting through the manuscripts that she was copyediting.

Finally giving in to her headaches, she'd find herself reaching for a family-size tub of pretzel rods she had stashed away in her desk in case of "emergencies." Munching mindlessly, she'd find herself easily finishing most of the tub before working her way through a single chapter. But she couldn't ignore that the carbs did bring her brain back online and allowed her to slog through the rest of her work. What Claire had no way of knowing was that it wasn't a lack of willpower on her part that created a certain midafternoon weakness for her old-fashioned pretzel rods—it was her genes. She felt guilty about her pretzel binges but uncertain about what else she could do to ensure a productive workday.

So I walked her through the first genetic self-test in the Eat for Your Genes pillar that I've designed (yes, I'll be doing that for you, too, soon enough) to find out her optimal level of carb intake.

"Instead of continually guessing your way diet after diet, wouldn't you love to know what your ideal carbohydrate intake is?" I asked her.

"Of course, who wouldn't!" she replied.

"Exactly. But you're going to have to follow my instructions as we go through some of the DNA Restart genetic self-tests. In doing them, you're going to discover some important things about yourself and your genes. So let's get started. I've brought all you're going to need for the first genetic self-test: a cracker and a timer."

"A cracker . . . ?" Claire smiled, thinking I was pulling her leg.

"No, I'm serious, and since you have no issues with gluten or wheat, we're going

---

[1] I have changed the names and identifying details in the case studies throughout this book to provide anonymity for individuals, patients, and their families.

to get started with this salt-free saltine cracker instead of the small piece of raw potato I use with all gluten-sensitive DNA Restarters," I explained.

"But I've probably eaten hundreds, if not thousands, of crackers over my lifetime," Claire said.

"Yes, but you've never eaten one with your genes in mind."

What Claire discovered through the DNA Restart Cracker Self-Test was that she had inherited multiple copies of a gene you'll be hearing more about soon called *AMY1*. Finding out her genetically based ideal daily carb consumption was key for Claire to be able to, for the first time in her life, eat for her genes. Most important for Claire, it wasn't just her problems concentrating that resolved, but she was also able to lose 14 pounds and very easily arrive at her ideal weight. For the first time in her life, Claire was eating for and not against her genes.

Now if you're wondering if you should be making the same changes that Claire made and increasing your daily carbohydrate consumption to match your DNA, you'll have to read on to find out. Your results from the DNA Restart genetic self-tests and the health recommendations found in the five pillars of this book will guide you to the diet that's just right for your very own genes.

Up until now, you've been essentially eating blind, without any personalized genetic wisdom to guide you. Our modern life is simply out of touch with our DNA. We've turned our backs on the wisdom contained within our three-billion-letter genetic code that our forebearers spent thousands of generations carefully annotating and preparing for us. As a physician and scientist, I spent the last 20 years researching the ways history, our genes, and the choices we make in our lifetimes intersect. I saw firsthand how trying to solve our modern health problems with a one-size-fits-all solution was simply not possible. Just like with Claire, one-size-fits-all dietary directives are not sustainable.

So building on my 2 decades of scientific research, I embarked on a 2-year journey, which took me across five of the world's continents, to bring together what I knew we needed genetically, nutritionally, and culturally that would lead to a practical and successful pathway to optimal living.

The more I traveled the more I explored how ancient methods of food production and preparation that our ancestors employed, such as fire and fermentation, played a decisive hand in reshaping the genes they subsequently passed down to us. Each one of our genetic ancestors faced the same challenge—survival. Every one of them had to find a way to survive immense odds, often when only scarcity prevailed, to find ingenious ways to procure, prepare, and produce food.

The results of each of these struggles are encoded within the genes you've inherited. A simple example is if you can enjoy dairy products as an adult, then it's a sign that your ancestors kept animals in order to drink their milk and gave you the genetics to do so as well. But as we've come to see with the explosion of the availability of dairy products worldwide and the dietary problems this has created, we didn't all inherit the same genetic knowledge.

My 20 years of research confirms that this deeper wisdom is not hidden and locked away deep within our bodies but can actually be made available to every one of us. I'm going to be your guide throughout this book, helping you unlock your DNA's hidden dietary rules that have been crafted specifically and only for you. Genetically speaking, there's no one else like you, so why should you be eating or living like anyone else? It's time for you to start taking control of your own genetic destiny and to change your life the DNA Restart way.

The DNA Restart is divided into five parts—one part for each pillar. In the 1st Pillar, Eat for Your Genes, you will be doing genetic self-tests that I've designed to get you eating the amounts of carbohydrates, fats, and proteins that are just right for your unique genes. In this pillar you'll also be taking a genetic self-test that will determine the specific amount of alcohol you should be consuming for your personalized optimal health. Depending upon the results of these genetic self-tests, you will be provided with specific, unique meal plans, recipes, and dietary guidelines for you to follow during your DNA Restart.

The 2nd Pillar, Reverse Aging, will be introducing you to ways in which you will be turning back the hands of your own genetic clock. To do this, I'm going to show you how to activate your body's own innate antiaging systems as well as prevent DNA aging damage in the first place, through both dietary and behavioral changes you'll be making during your DNA Restart.

Umami—the Japanese word for deliciousness—is the fifth human taste. It's also going to be the crucial weapon in your weight-loss arsenal for the next 28 days. In the 3rd Pillar, Eat Umami, I will show you my practical, umami-based dietary strategies that will leave you feeling much more satisfied naturally, with a lot less food, both during a meal and long after you've finished eating. Using strategies that I developed after meeting with chefs of Michelin three-star restaurants around the world who expertly use the principles of deliciousness or umami every day, I've created food preparation techniques that are not only delicious but naturally slimming as well.

In the 4th Pillar, Drink Oolong Tea, you'll discover the powerful benefits associated with drinking oolong tea every day. This special tea has the potential to shift your microbiome in a beneficial direction—by promoting microbes that favor health over obesity. Drinking oolong the DNA Restart way also has the added benefit of stopping you from absorbing fat from your diet, as well as specifically targeting visceral fat, or the proverbial "belly-fat" weight. In this pillar you'll also learn the best ways to prepare oolong tea that will allow you to reduce oxidative stress without adding any additional calories during your 28-day DNA Restart.

In today's modern, fast-paced world, being continuously sleep deprived, having to always eat quickly, and spending too much time away from the ones we love has become the norm. And all the stressors created by modern living have taken a serious toll on your body and, most importantly, on your DNA. The 5th Pillar, Slow Living, is the real linchpin of the DNA Restart because it will teach you the secrets behind recalibrating your life to maximize your genetic potential. Continuous stress puts all of us in an unhealthy state that will even change the DNA we pass on to the next generation. In this pillar I've included the essential exercises that I've specifically developed for the DNA Restart to help you tackle the modern problem of too many choices, way too many available calories, and not enough time.

Some of the changes you're going to be making in the DNA Restart will not be easy. That's because you'll be reprogramming not only the way you behave but how your genes do as well. That's why the final section of the book is the DNA Restart Road Map to Optimal Health and Longevity, which has been designed to get you to your goals, step by step. The road map distills all of the five pillars and includes the meal plans, recipes, and exercise regimens that you will need to follow to lose weight and bring yourself back to an ideal state of genetic health. This is the same plan that I followed to lose more than 30 pounds and turn back my own genetic clock. But the only real way to feel the effects of the DNA Restart is to start. So let's begin.

PART I

THE DNA RESTART
1st Pillar

Eat for Your Genes

The DNA Restart is a completely new way for you to view your genetic inheritance. The 1st Pillar, Eat for Your Genes, will help launch your 28-day DNA Restart plan to completely transform your relationship with food and, most importantly, your relationship with your genes.

But let's be up front: It's not going to be easy.

As I'm sure you've discovered for yourself, most people who've lost weight by dieting don't keep the weight off over the long term. As a physician and scientist, I know that most diets fail because of two important flaws. The first is simply a mind-numbing, restrictive lack of a variety of food and meal choices, making it impossible to stick to the diet in the long term.

The second and most important reason is that, until now, there hasn't been a single diet that is designed with every single person on this planet in mind. Deep down, modesty aside, you know full well that you are not like anyone else. Nor have you ever been. What this means on a genetic level is that although you may be very similar to other people, there's absolutely no one else exactly like you in the entire universe.[1]

Not even close.

Far from being benign, eating for someone else's genes can be deadly. Take Thomas, for example. During high school and college he both ran and swam competitively and had no problems keeping his weight in check. But later, when he became a busy father with two young kids at home and a demanding professional

---

[1] As deep genetic sequencing revealed, even monozygotic, or "identical," twins do not always have the exact same DNA. That's without even accounting for a myriad of inherited and evolving epigenetic differences.

1

life that saw him traveling often, he found that he had little time left over for personal self-care like exercise. He rarely, if ever, donned his sneakers or found the time to hit the pool for an early morning swim. This lack of exercise combined with still eating the way he did in college left Thomas with an ever-increasing waistline. At the age of 47, Thomas now found himself the not-so-proud owner of a dreaded and bloated midlife beer belly.

After struggling to get into his business suit one morning, and being barely able to button his pants, Thomas decided on the spot that he'd had enough. It was time for him to make a change. He knew that if he didn't make some serious lifestyle modifications soon, he'd end up just like his two brothers, who both had diabetes and were in the early stages of heart disease.

With newfound motivation, Thomas was able to tap into the same dedication and physical persistence he had shown during his athletic youth and began to find his stride. With the help of a personal trainer, Thomas started off slowly at first. Within a year, he was back at the gym almost daily, both swimming and running six times a week. He also made significant dietary changes that included cutting out all cereal grains and substantially increasing his intake of protein. It took another year after that, but Thomas was now back at a level of physical performance he couldn't have dreamed of 2 years earlier. And one of the added benefits was that his love handles and dreaded beer belly were now a thing of the past. In their place was a peak athlete who now fixed his sights on qualifying for the Ironman World Championship competition in Hawaii just before his 50th birthday.

That's when Thomas's energy levels unexpectedly started to change. It was just barely perceptible in the beginning; his wife was the first to notice that he wasn't able to jump out of bed for his early morning swim with his usual vigor. This was followed by increasing feelings of fatigue no matter how much he slept. Within 6 months Thomas found himself completely and thoroughly physically exhausted without any good explanation. Maybe he was training too hard, or maybe his body just wasn't up for the challenge. When he found himself consistently unable to get up before his family for his early morning run, Thomas decided it was finally time to seek some medical help and made an appointment to see his doctor. After getting a clean bill of health from his physician, Thomas was at a loss for what to do next. Not having the energy to train any longer, he found himself becoming more and more depressed, his hopes of qualifying for the Ironman World Championship fading with each passing week.

What Thomas could not have known at the time was that his dietary changes were singularly responsible for his deteriorated health status. It was only after he took the leap of faith and did the DNA Restart that he discovered why he was feeling so ill. The reason was simple: No one, including his trainer who had recommended the high-protein diet in the first place, was thinking about Thomas's DNA. If they had, they would have discovered that by eating more protein in the form of red meat, Thomas was slowly but surely rusting to death.

The reason that Thomas became sick, as I was to discover and explain to him later, was not at all his fault. The answer was to be found in his genes. That's because Thomas unknowingly inherited versions of a gene called *HFE* that result in a condition called hereditary hemochromatosis.

People with this condition absorb way too much iron from their diets. In Thomas's case, the condition was made worse when he began to eat even more red meat, causing him to eventually literally rust from the inside out. All of that extra dietary iron builds up in the organs of people with hereditary hemochromatosis. Unchecked, this internal rusting can lead to things like cancer, heart disease, and diabetes. Thomas discovered the reason for his health deterioration through the DNA Restart, which gave him the knowledge to start eating for his genes. He also learned that his hemochromatosis wasn't the only reason his health was suffering. Through the DNA Restart genetic self-tests, he discovered that he actually had the genetic capacity to eat more carbohydrates. He has now significantly reduced his intake of red meat while reintroducing whole cereal grains and legumes, and the results speak for themselves. This increased carb allowance helped fuel his rigorous daily training program and road to recovery—a big surprise to both Thomas and his trainer. Thomas's energy levels are fully restored, and he let me know recently that he was able to not only qualify for but complete his first Ironman World Championship event. Not a small feat.

Up until now, there never has been a bespoke diet created and tailored for only your unique genetic needs. You have actually never once intentionally and methodically eaten for your genes.

That's about to change. I've created signature DNA Restart Self-Tests that are designed to hit upon the highest-yielding scientifically based results, and you can do these tests easily at home. And just like the multitude of DNA Restarters who have experienced the power of eating for their genes by employing the DNA Restart Self-Tests, once you start eating for your genes, you'll never want to go back to eating genetically blind again.

Another added benefit of the DNA Restart self-tests is that you control what to do with the results. Most people are not aware that there is currently no universal US federal protection against genetic discrimination. This means that every time you subject yourself to a genetic test with a third party, there's no guarantee that your genetic information or results will remain private. The Genetic Information and Nondiscrimination Act (GINA) is limited in the types of protection it can provide you. Many people are shocked to find out that they may have absolutely no protection from genetic discrimination in matters of disability and life insurance. Some US states have been seeking to fill in the discrimination gaps left open by GINA, but there remains much work to be done. Please carefully consider this and keep it in mind when you consider any genetic testing done with a third party.

This is why I've created genetic self-tests that you can do at home without putting yourself at risk for future discrimination. Your genes belong to you, and your genetic information should not be available for others to access without your permission.

CHAPTER 1

Why You Should Eat for Your Very Own Genes

The first and most important genetic self-test that I'm going to introduce you to is an incredibly powerful tool that will allow you to individualize your carbohydrate intake levels. The results of this first self-test will help you find out which one of three carbohydrate consumption categories you fall into.

Until recently it was assumed that we all inherited a copy of each gene, one from each parent. And that's why it was thought that you have two copies of every gene. Oh, how wrong we were. It turns out that some of us have a little more or less DNA than others. And far from being insignificant, the number of copies of genes you've inherited can have a tremendous impact on your life and health.

To get more nutritional horsepower out of your genome, for example, an ancestor of yours might have gifted you with multiple copies of a certain gene. Instead of the usual two copies of a gene, you may have inherited even a dozen copies or more. The technical term for this phenomenon is *copy number variation*, or CNV, and it seems that this has happened at many different times and places in our evolutionary history.[1] We all vary in the number of some of the genes that we have inherited—even within the same family.

Okay, so why should you care today how many copies of a gene you may have inherited from a specific ancestor thousands of years ago?

Because many of these variations in the number of specific genes you've inherited were passed on from your ancestors as an advantage when eating certain things for breakfast, lunch, or dinner. That dietary advantage for the specific nutritional environment would then be passed on and maintained down a specific genetic ancestral line. And as you're going to discover yourself, the copy number of certain genes you inherit plays a very important role in determining your optimal diet today. Using this information, you can follow an individualized diet that will help you get

---

[1]   You can learn more about the impact that CNVs can have on your health by reading one of my previous books, *Inheritance: How Our Genes Change Our Lives—and Our Lives Change Our Genes.*

to and stay at your ideal weight, painlessly. And most importantly, it will do this while improving your overall health and increasing your genetic longevity. You'll gain all of these benefits simply because you understand your unique CNV when it comes to certain genes and how that should dictate your dietary choices. See why you should care?

That's why for the most part doing what others do when it comes to diet and lifestyle may be perfectly fine for them, but over a lifetime of bad genetic decision making, the consequences can even turn deadly for you.

Since the first complete draft of the human genome was published back in 2001, we are becoming more aware that the most striking and impactful genetic differences between us are actually found in genes that impact our diet. The way our bodies digest and use energy from the food we eat depends more upon what your recent genetic ancestors ate than we ever imagined. And these genetic differences can have an incredibly powerful influence on not just our basic nutritional needs but also on the amount of carbohydrates we can eat, down to the ability for us to either thrive or wilt upon consuming a large amount of protein.

To illustrate the importance of how genetic differences between people can affect what they should be eating, I have often asked my patients what type of fuel they use to power their cars. Here's why: Some makes and models require a high-grade gasoline while others can get by with the lowest and cheapest type around. Others even require a special type of fuel such as diesel or propane, while some cars are now powered by electricity alone. When it comes to your car, you find out which type of fuel to use by consulting your vehicle's owner's manual.

Now you might think that we weren't born with a similar type of instructional manual. Or worse, you think or have been told that a manual like that might exist and that it would be applicable to *all* of us. In fact, what you have within you is incredibly more detailed and impressively individualized. I'm talking about a three-billion-letter genetic code that is full of individualized and unique wisdom that was collected and annotated over millennia *just for you*. Every nutritional adaptation that allowed your genetic ancestor to survive long enough to pass on that knowledge to his or her own children is in there—a veritable genetic tapestry gifted to you from every direct genetic ancestor you have ever had.

That tapestry, when it's spread out, is very, very large: about three billion genetic letters or nucleotides' worth making up your genome. And you actually have two genomes—one from your mother and one from your father, which is why you're not identical to either one. Within our genomes we also each have somewhere in the ballpark of 20,000 genes that do most of the genetic heavy lifting in our bodies. Nearly every single one of your trillion cells has an entire copy of your genome inside it.[2] That's how the folks on *CSI* were able to identify people with only a very small sample of hair or tissue.

Most of the knowledge within your own DNA is on genetic autopilot, requiring little conscious input from you as your cells use it on a moment-to-moment basis to keep you alive. But if you really want to start to eat for your genes, you're going to have to learn to speak their language. That's why I've devised a few crucial at-home self-tests in the DNA Restart that will help you start decoding your own unique genetic inheritance. This information is required so that you can eat for a better, longer, healthier life.

I've also included a few case studies of patients and colleagues to illustrate the power of what can happen when you finally start to eat for your own genes and stop eating for everyone else's. This is your life. Let's get you started on your way to eating for your genes.

---

[2] The only cells that don't have DNA are mature red blood cells. They actually got rid of their DNA to make more room so they could pack themselves full of more hemoglobin.

CHAPTER 2

Why Some People Can Thrive on Carbs: The Saliva-Carb Connection

You may not pay much attention to your saliva, but it's a veritable cocktail of proteins and enzymes that have been fine-tuned to begin the process of digestion long before your food hits your belly. You can think of enzymes that are involved in digestion like kitchen appliances. Just like some people have kitchens without any appliances while other people have ones that are decked with the latest high-power culinary gadgets, we all differ in the types of genetic tools found in our saliva and that we have inherited to help us break down and digest our food.

Most people have an enzyme within their saliva called amylase, which, like a giant pair of shearing scissors, has the ability to cut apart big and bulky starch molecules into simpler sugars. That's the first step in making big threads of starch molecules available for the body to use easily as energy. And amylase is really good at cutting up starch. Don't believe me? If you're curious to see just how powerful amylase is, add your own saliva to a full jar of a starch-rich baby food like pureed bananas and put it back in the fridge. By the morning, the entire jar should be liquefied.

Some of us have supercharged saliva that's just waiting to cut apart the carbohydrates we eat by being packed full of amylase (saliva that's turbocharged can have up to 50 times more amylase). As it turns out, researchers were surprised to discover that we're not all endowed with the same amount of amylase in our saliva. Some people were actually found not to have any at all!

Remember what I told you about copy number variation? Well, you may or may not have inherited multiple copies of the gene your body uses to make the protein amylase, called *AMY1*. And the more copies of *AMY1* you've inherited from your parents, the more amylase you have in your saliva right now.

There are three things that really surprised researchers about the *AMY1* gene. The first was that there's an incredible degree of variability when it comes to how many copies of *AMY1* people inherited. Some people have none, while others have as many as 20! That's copy number variation for you: If you have multiple copies of amylase, you are lucky enough to speedily burn through and digest a tremendous amount of starch while still only chewing your food.

Have no copies of *AMY1*? Well, your saliva will suffer for it because it will contain no amylase, making the task of breaking down carbohydrates metabolically daunting for your body.

Sometimes genetics can be that simple.

A similar genetic evolution is behind skin color: The darker your skin, the more your ancestors needed to be protected from the DNA-damaging radiation of the blaring overhead sun. And so it turns out, just like the color of your skin, the number of *AMY1* genes you've inherited is not random either.

It is actually highly dependent upon *where* your own unique genetic ancestors hail from. Okay, let me simplify all that: If you come from ancestors who relied heavily on starches, such as farmers growing and consuming cereal grains, you'll likely have been gifted with multiple copies of *AMY1* that can make tons of amylase in your saliva. If your recent genetic ancestors, on the other hand, were more into meat than potatoes, then the number of amylase genes you've inherited from them— not so much.

This is why one of the biggest genetic differences between you and your best friend will likely be in the genes that are in some way involved in the foods you should or should not be eating. Likewise, that's why the levels of amylase in your saliva will not be the same as someone else's with whom you may be sharing a meal. The more starch your ancestors ate in their daily diet over generations, the more copies of *AMY1* genes you've inherited. It's as simple as that. And these genetic differences and the duplicating of the *AMY1* in some of our ancestors I believe really kicked off with the domestication of grains in the last 10,000 years. There's good reason for this, since a historical shift to a higher intake of carbohydrates through the consumption of cereal grains would be aided, as I'm about to show you, by having more copies of the amylase-producing *AMY1* gene.

If your ancestors switched to eating more cereal grains such as wheat and rice, you're much more likely to have inherited multiple copies of the *AMY1* gene with every generation. That's just the way genetics works. But you may be wondering: If it's useful to have multiple copies of *AMY1*, why didn't we all evolve to have them?

The answer has to do with what I like to call biological home economics. Imagine that your genes are like employees who require a salary to work. The more copies of a gene you inherit, the more you will have to pay them to have them work for you. But if there's not much starch-digesting work to do because your ancestor is not spending his time eating bread and rice every day, then there's not much starch for the amylase from the *AMY1* gene to break down. So what then? The way biology works is that it would rather you not have to go and needlessly expand your genetic workforce with more copies of *AMY1* genes and instead save all of that biological energy to spend on other important physiological functions.

If all this wasn't enough for you to want to jump right in and find out how many copies of *AMY1* you might have inherited and how much amylase you correspondingly have in your own saliva, there's still one significant thing you need to know.

And here's the real kicker about amylase. You'd think that if you have saliva overflowing with amylase from additional copies of *AMY1*, then a carbohydrate meal would send you soaring into a sugar rush high because all that starch gets almost instantly converted to more sugars such as maltose (which is a disaccharide on its way to eventually becoming a simpler sugar like glucose). If that were the case, this would mean that people who have many copies of *AMY1* would not fare well eating a lot of carbohydrates because they'd be digesting starches much faster, which would send them barreling down a road to eventually higher risk for insulin resistance and obesity.

As it turns out, your body is more clever and interesting than anyone had ever imagined. As scientists from Monell Chemical Senses Center in Philadelphia together with colleagues from Rutgers University in New Brunswick, New Jersey, discovered to their surprise, the reverse actually happened: People with more amylase because of multiple copies of the *AMY1* gene had *lower* levels of glucose than people with fewer copies of *AMY1*, who had *higher* levels of glucose. On the surface, of course, this doesn't make much sense, since people with more amylase are better and faster at digesting starches, which would mean that their glucose levels should be spiking.

But what they saw was the exact opposite.

Here's what made their experiment really clever: They also decided to measure the participants' insulin levels. Insulin is the hormone that helps shuttle glucose into cells so that they can use it for energy. And that's how the researchers got their answer. The people with more copies of *AMY1* were much quicker and more responsive to produce insulin in anticipation of the coming flood of sugars. So they were much better prepared—kind of like when Amazon hires more workers in anticipation of the yearly online holiday crush.

I believe that eating out of sync with our genes is the reason why some people are more prone to developing obesity and insulin resistance than others on a diet that's relatively high in starch/carbohydrates. In fact, research is now beginning to back this exact view, as it's been found that people with the *lowest* copy number of the *AMY1* gene are actually much more likely to be obese.

When you take the long view—the one that biology loves to take—then this is completely in line with what you would expect. After all, when it comes to survival, having the ability to break down and deal with sugars released from eating a pre-

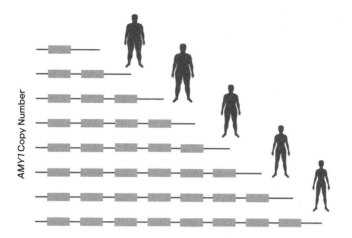

The lower the number of *AMY1* genes (each one represented here as a shaded box) you've inherited from your parents, the more likely you are to be obese when eating a carb-heavy diet.

dominantly carbohydrate-based diet is a good thing for people who eat that kind of diet on a regular basis.

The downside, of course, is when you didn't inherit many copies of the *AMY1* gene and still try to eat a lot of carbs. Very likely you will end up obese or diabetic or even both. So knowing how much amylase is in your saliva can be incredibly powerful to get you eating just the right amount of carbs. Yet be warned that no matter what genes you've inherited, if you're exposing yourself or your family to products that are needlessly loaded with carbs (think of the entire processed food section of your local supermarket), your waistline will eventually become thoroughly ravaged no matter how many copies of *AMY1* you've inherited.

Honey is the one shining example of a simple carbohydrate sweetener that you are encouraged to enjoy on the DNA Restart. Filled to the brim with phytochemicals and antioxidants, honey was also the first antibiotic ointment used by your ancestors, who recognized its microbe-killing potential 4,000 years before the discovery of penicillin. I've spent many years looking for new microbe-killing compounds in different varieties of honey. Some of these honey-derived compounds, such as methyl syringate and methylglyoxal, have been shown to kill bacteria such as *Helicobacter pylori*, which can cause duodenal ulcers and even gastric cancer. Honey has even been shown to kill antibiotic-resistant "superbug" bacteria such as methicillin-resistant *Staphylococcus aureus*, or MRSA. For all these reasons and more, you can have a maximum of 2 teaspoons of honey every day during your 28-day DNA Restart. When purchasing honey, look for some that hasn't been heated or filtered, and choose the one that appears the most viscous. If you don't mind a stronger taste, look for a darker-colored honey. As you'll see later as well, I've incorporated honey into some of the DNA Restart Recipes.

Unlike honey, processed sugars aren't just empty calories; they lack essential antioxidant and antimicrobial phytonutrients, which are crucial to sustain and nurture your body and DNA. That's why irrespective of your *AMY1* gene copy number, all processed sugars are banned on the DNA Restart.

NO MORE SOFT DRINKS EVER

If you're looking to gain a lot of weight and make yourself insulin resistant at the same time, make sure to consume as many regular soft drinks as you can. One of the most common causes of obesity in my patients was from the overconsumption of liquid calories. What a waste! This phenomenon, of course, was not limited to my practice but has been chronicled as one of the leading vanguards in the obesity epidemic.

Obviously, drinking empty calories is not a good idea. And if you think you're doing yourself a really big favor by having a diet soda now and then, think again. As a society, we are consuming a convoy of tanker trucks full of artificial sweeteners every year. And we have absolutely no idea what the long-term health effects are of a lifetime of consuming this amount of sweeteners.

Or do we?

Studies are starting to trickle in that I and many of my scientific colleagues believe will turn into a flood of anything but sweet conclusions concerning the long-term health risks and consequences of consuming both sugar and artificially sweet-

DNA Restart Health Tip #1
You are welcome to enjoy a maximum of 2 teaspoons of honey every day. No processed sugar consumption allowed.

Honey is not just another form of sugar. It contains a very special combination of phytochemicals (which includes powerful antimicrobials and antioxidants) derived from all of the millions of flowers and trees honeybees needed to visit to make it. Honeybees need to gather the nectar from 35,000 flowers to make every teaspoon of honey!

Lost 15 pounds during my DNA Restart without even trying that hard! Finding out that my Carb Intake Category was Restricted kept me on track over the 28 days when planning my meals.

—Natalie, 39

ened beverages, which also include coffee and tea. As for sugary drinks, by now it's fairly clear that consuming them in excess is in fact linked to developing Type 2 diabetes. But hold on a minute, because this is where things get interesting.

If you happen to be following the advice of numerous health agencies and replacing your sugar-laden drinks with their diet equivalent, you probably think that these "diet" drinks are going to protect you from developing diabetes. Well, researchers at the University of Cambridge recently published findings from a study that addressed this very issue by following the habits of more than 25,000 people in the United Kingdom over a 4-year period. What they found was that even if you do replace sugar-sweetened drinks like soft drinks and tea and coffee with artificially sweetened ones, you still do not lower your overall risk of developing Type 2 diabetes. In fact, the only way to really lower your risk of developing Type 2 diabetes is to switch to drinking unsweetened coffee or tea or having good old plain water. Why this is the case is still not clear.

What I think is happening is that consuming artificial sweeteners is taxing your metabolic system. Just like your taste buds, your body is being fooled by a sweet taste that never materializes as sugar in any of its forms. So over years of consumption, it's possible that your body stops listening to sweet hormonal signals from artificial sweeteners. After they've experienced so much rejection and disappointment, who can blame your cells for giving up and stubbornly becoming insulin resistant, the hallmark of Type 2 diabetes?

If you're truly keen on avoiding untold risks to your health, what alternatives do you really have? Maybe I'm a little more cautious than some health agencies out there. But why take part in this research, when you're totally free to opt out?

That's why on the DNA Restart you are expressly barred from the consumption of any artificial sweeteners. But I'm not just going to leave your active beverage life totally high and dry, because Pillar 4 on the DNA Restart is all about a health-enhancing alternative drink that will change your waistline, microbiome, and much more.

You'd have to be hiding under a rock to not be aware of the dangers associated with things such as sugar-sweetened beverages like soft drinks, fruit juices, and even

R̸ DNA *Restart Health Tip* #2

No sugar substitutes–whatsoever! Here's an easy list to watch out for.

1. Acesulfame K
2. Aspartame
3. Cyclamate
4. Neotame

5. Saccharin
6. Stevia
7. Sucralose
8. Sugar alcohols, such as xylitol

---

chocolate milk. So not surprisingly, none of those extraneous sugars are allowed on the DNA Restart. But don't even think about this now, as I'll be delving into what you can and cannot have on the DNA Restart in greater detail in the 2nd Pillar.

There are very good reasons for some of the vilification of having too many carbohydrates in our diets. Most of the carbs widely available today in their hyper-processed forms would be unrecognizable to *all* of our genetic ancestors, but they don't seem to be affecting all of us in the same ways. This is another important reason why you should be eating for your genes. Why should you be eating carbs like anyone else? It's time that you start eating carbs with your *AMY1* genes as your guide.

So let's get practical! What I've designed with the DNA Restart Cracker Self-Test is a way to finally eat the amount of carbohydrates that's in line with the number of *AMY1* genes you've inherited. The results from this self-test will place you in one of the three Carbohydrate Consumption Categories. After you complete the self-test, I will be giving you the individualized meal plans, including a few recipes and eating advice that will be tailored for your very own DNA Restart.

Let's get your carb intake aligned with your genes!

CHAPTER 3

The DNA Restart Cracker Self-Test

In this pillar, Eat for Your Genes, we're going to be unpacking key parts of your own genome at home with a few genetic self-tests, which I've designed just for you. The first genetic self-test will be the DNA Restart Cracker Self-Test. Results from this experiment will indicate how much amylase you have in your saliva and, through that result, about how many copies of the AMY1 gene you've inherited from your parents.

What You'll Need for the DNA Restart Cracker Self-Test:

1. One saltine cracker (must be unsalted) or, if you're following a gluten-free diet, a dime-size piece of raw peeled potato

2. A timer

3. A pen or electronic device for your note taking

DNA RESTART CRACKER SELF-TEST: LET'S DO IT!

Get your cracker ready (or if you're gluten-free, a dime-size piece of raw peeled potato) as well as a timer and something to take notes with. Make sure that your saltine cracker is unsalted, as the self-test requires it. Now break the cracker approximately in half, or if you're using a piece of raw potato, have it ready and place either one in front of you. The goal of the DNA Restart Cracker Self-Test is to find out which of the three Carbohydrate Consumption Categories you fall into: Full, Moderate, or Restricted. You will get this information by the amount of time it takes for a change in taste to occur when you're chewing either the saltine cracker or potato. The longer you're chewing, the more likely the taste will change. If you never detect a change in taste, that's normal (and significant!), too. To ensure that you get the most accurate results, you'll be running through the experiment three times and averaging the results.

Now get your timer out (most phones have at least one app for this function). Whenever you're ready, place the cracker or potato in your mouth and start timing and chewing. You'll need to pay really close attention now, as the starch in the cracker or potato may already be starting to be digested by amylase in your saliva. If you feel yourself wanting to swallow, that's perfectly normal, but try to stop yourself. Try to imagine that you're simply chewing a piece of gum as you keep chewing.

As soon as you detect a change in taste or if you reach 30 seconds while timing, stop chewing, swallow, and note the time. Rerun the same self-test two more times. Now I want you to add up the three times and divide by three so that you can get to your DNA Restart Carbohydrate Consumption Category. Take your combined score and have a look at the following table to find your personal Carbohydrate Consumption Category.

The DNA Restart Cracker Self-Test:
Carbohydrate Consumption Categories

Time in Seconds for Taste Change	Carbohydrate Consumption Category
0–14	Full
15–30	Moderate
More than 30	Restricted

It's important to note that this will be your guide for how much carbohydrate you should be having throughout your 28-day DNA Restart. From here you go to the Carbohydrate Consumption Estimate Guide, which will give you the specifics of your daily carb cost as a percentage and in grams.

The DNA Restart Carbohydrate Consumption Estimate Guide

Your Carbohydrate Consumption Category	Carbohydrate Intake* for Women in Grams	Carbohydrate Intake** for Men in Grams	Percentage of Carbohydrate Intake	Carb Cost Allowance in Points
Full	250	325	50%	13–16
Moderate	175	230	35%	9–12
Restricted	125	165	25%	5–8

*This is assuming an approximate average caloric intake of 2,000 kilocalories per day.

**This is assuming an approximate average caloric intake of 2,600 kilocalories per day.

Remember that these numbers are meant purely as an estimated guide, since your focus for the next 28 days will be to bring your life in line with your DNA and not to count calories. That being said, if you really want the weight to come off, you're going to have to be diligent regarding the amount and quality of food you eat. Today, we are all consuming more food than our DNA thrives on. Overeating literally stresses the body, which hurts our genes, saps our youth, and reduces our longevity. To make matters worse, our food has been stripped of essential phytonutrients and minerals, which used to nourish and strengthen our DNA. Pillars 1 through 5 of the DNA Restart have been designed to bring your life back in line with your genes.

Now if you're a visual type of person, I've also provided you with the pie charts below to give you the dietary breakdown for fats, proteins, and carbs.

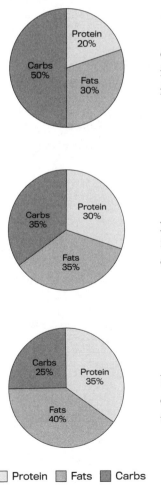

If your DNA Restart Carb Category is Full, you can have up to 50 percent of your calories coming from carbs, 20 percent from protein, and 30 percent from fats.

If your DNA Restart Carb Category is Moderate, you can have up to 35 percent of your calories coming from carbs, 30 percent from protein, and 35 percent from fats.

If your DNA Restart Carb Category came out to be Restricted, you can have up to 25 percent of your calories coming from carbs, 35 percent from protein, and 40 percent from fats.

☐ Protein ☐ Fats ☐ Carbs

> *I've been struggling with Type 2 diabetes for more than 10 years. I got my diabetes under so much better control using the results from my DNA Restart Self-Tests and DNA Restart carb cost system guide. My doctor couldn't believe it and neither could I.*
>
> —Ryan, 64

I've also created a DNA Restart Carb Cost Allowance Guide so that you can effectively keep track and stay on top of your weekly carb dietary intake (see Chapter 37). The list I've provided has both common cereal grains and a few vegetables and their corresponding carb cost. The carb cost allowance system is meant to be a guide to get you to eat more in line with your DNA as per your results from your self-test; it is not meant to be a definitive list of all the food that may contain carbohydrates.

Here's how to calculate your weekly carb cost allowance:

1. Take the DNA Restart Cracker Self-Test on page 15.

2. Find your Carbohydrate Consumption Category from the results of your cracker self-test on page 16.

3. Your carb cost is like your weekly allowance for you to buy the carbs you need. Just make sure that you stay within your allowance budget for the week!

Carb Cost Allowance Breakdown:

Full Carbs: 13 to 16 points of carbs per week

Moderate Carbs: 9 to 12 points of carbs per week

Restricted Carbs: 5 to 8 points of carbs per week

CHAPTER 4

What Happens When You Don't Eat for Your Genes?

Fiona was in her early thirties and desperately trying to get pregnant. She had never really thought seriously about starting a family, but all that changed for her when she met Will during her training in clinical psychology at the University of Pennsylvania. Most of the last decade and a half was devoted to working to complete her training, but now, as a newly minted psychologist, she found a blossoming desire to have a child with Will, who had finished his law degree 2 years prior.

For the most part Fiona was always healthy, with the occasional bout of stomach pains that oscillated between loose stools and constipation that she attributed to the stress from the rigors of her program. And so they tried to conceive naturally at first, but with no success.

They often laughed at the irony that after many years of working so hard to avoid getting pregnant, now the tables had turned and they were doubling their efforts in their attempt to have a child. The first year was up, and with not even a small sign of a pregnancy in the making, they sought some medical advice from a local reproductive doctor. After making an appointment and going through the initial medical screening—a sperm count for Will and blood tests for both—they were scheduled to see the doctor and get the results.

The news for the most part was good. Their doctor said that all the tests came back normal, besides a little microcytic anemia and low ferritin on Fiona's part, which can be common in menstruating women of her age. She suggested that Fiona add an iron supplement to her diet or see a nutritionist they had on staff to get some nutritional advice. Other than that, there wasn't anything of real concern.

The doctor then discussed their options, including whether they still wanted to keep trying on their own. Perhaps they would increase their odds at a successful pregnancy by timing their sex with Fiona's ovulation. This sounded like the most reasonable option at the time for Fiona and Will. The doctor ended the visit with a gentle reminder that once some women reach their midthirties, their odds for a natural pregnancy start to decline precipitously. So they shouldn't spend too much time contemplating other options, as time wasn't on their side.

As her doctor had advised, Fiona made a follow-up appointment with the nutritionist on staff and spent more than an hour filling out a very exhaustive dietary questionnaire. She was happy to have seen the nutritionist, because she actually confirmed that Fiona was eating a well-balanced diet. The only real suggestions were, given Fiona's anemia, to try to increase her dietary intake of iron and definitely increase her fiber consumption from whole cereal grains to help her become a little more regular.

Before they knew it, another year passed for Fiona and Will. They agreed that after trying for this long, they should return to the reproductive health clinic.

That's where I came in. Will e-mailed me that evening to see if I had any other suggestions or advice for them to maximize their chances of a successful pregnancy, and we scheduled a time to speak the next week. I first met him a few years prior when he was in New York City working at a small law firm. We spent a few minutes catching up, and then he let me know why he was calling. I mentioned to Will that there was some genetic testing they could consider, but for the most part their reproductive clinic had a great reputation, and the advice they were being given was medically sound. I asked Will a few more questions about Fiona's health.

"You know, ever since she changed her diet, her anemia hasn't improved at all since the last time she checked, and she's been having much more stomach issues than I remember her having in the past. I thought that might be because of, you know, all the stress of trying to get pregnant."

I reflected quietly on everything Will had just shared with me for a moment before asking, "Will, does she eat gluten?"

"Yeah, of course, she loves everything with gluten," replied Will.

This led to a lengthy conversation between us about the little-known silent symptom of celiac disease, and then some follow-up testing on Fiona, which conclusively proved that she was, in fact, fully gluten intolerant.

What's interesting is that with the rise in public awareness of celiac disease, many more people are aware of the havoc gluten can cause your gut if you have the disease, but they have no idea that it can cause serious fertility issues for those suffering without a proper diagnosis and subsequent dietary changes. Another little-known fact is that celiac disease can cause anemia, like it did for Fiona, which had nothing to do with the amount of iron she was getting in her diet but rather the intestinal changes that happened because of celiac disease. Without a diagnosis, there would just be no way of knowing that those daily iron supplements weren't actually being absorbed and used by her body.

The good news about Fiona and Will is that with the proper diagnosis of celiac disease, which included an intestinal biopsy, and the avoidance of all things gluten, it was possible to become fertile once again. And that's just what Fiona and Will happily discovered; they're now expecting their second child.

Classical cases of celiac disease are thought to have a genetic component. Most people with celiac disease have some type of shared western European ancestry with two genes in particular, *HLA-DQA1* and *HLA-DQB1*, figuring as prime suspects. But not everyone with those or the multitude of other genes that have been implicated go on to develop celiac disease. There are screening tests for celiac disease—the most commonly available is the immunoglobulin A tissue transglutaminase, or tTG-IgA test—but for this type of testing to be meaningful, ironically, you must be eating lots of gluten.

The way to a diagnosis is usually from a tissue biopsy, where a piece of tissue is taken directly from a patient's small intestine, but as I said, you need to be eating gluten. Even with today's increased awareness of celiac disease, many people just like Fiona regrettably still go undiagnosed. Part of the reason may be the chameleon-like nature of the condition's symptoms, which can include everything from bloating to iron deficiency anemia, joint pain, fatigue, anxiety, and even infertility, like we saw with Fiona.

So you might be thinking by now that as part of the 28-day DNA Restart plan, I'm going to have you go gluten-free.

The simple answer to that is no.

Though gluten is much maligned, and rightfully so in people with celiac disease, for the most part I believe that when gluten-tolerant people throw out all things "gluten," we are excluding a great source of nutrition from our daily dietary life (more on this in Pillar 2: Reverse Aging). This is especially true when you consider what I told you about the evolution of duplications of the amylase-producing gene, *AMY1*. There's no doubt some of us have inherited genes that allow us today to safely and healthily eat more carbohydrates.

These carbohydrates can come from many sources—including whole cereal grains—some of which contain gluten. But not all cereal grains contain gluten. Rice, amaranth, and quinoa, for example, are naturally gluten-free. The most important question when it comes to gluten is, why are so many people finding themselves sensitive today?

HISTORY OF CELIAC DISEASE

Celiac disease is anything but a recent condition. That probably makes a lot of sense, since experts believe it was more than 10,000 years ago that cereal crops such as barley and wheat were first domesticated. And that's a long time when it comes to diet and genetic adaptation.

Many of the crops first grown in the area aptly referred to as the Fertile Crescent helped provide the caloric energy to fuel many of the world's great ancient civilizations, such as the Babylonian and Assyrian empires. This region spanned an incredibly large area in what is today modern Iraq, through to Israel, and all the way to the Nile delta, which is in modern day Egypt.

The first recorded case of celiac disease was described by a Greco-Roman physician named Aretaeus of Cappadocia, who lived about 1,800 years ago in what is now Turkey. Aretaeus named the condition *koiliakos*, a Greek word, meant to describe the abdominal suffering that his patients were feeling, and what millions of people since who suffer from celiac disease would unfortunately understand really well.[1]

Today we consider celiac disease to be an autoimmune disease that is thought to affect around 1 percent of the world's population. Yet it took all the way until the 20th century for an actual celiac-gluten connection to be established. In 1950, Willem Dicke published his doctoral thesis describing how when afflicted patients were given strict diets, and stopped eating foods that contained gluten such as wheat and rye, their symptoms would improve.

We still do not fully understand why some people develop celiac disease. What we do know is that depending upon where your ancestors hail from—western Europe seems to be a genetic hot spot—you might be at a higher risk for celiac disease. As I told Fiona, there are many genes that have been associated with celiac disease, with two genes in particular, *HLA-DQA1* and *HLA-DQB1*, as prime suspects. However, not everyone with these genes goes on to develop celiac disease, and it seems that there might be environmental factors that trigger the condition when a susceptible person is exposed to gluten.

R℞ *DNA Restart Health Tip #3*
Here's a list of cereal grains that contain gluten.

1. Barley	7. Oats (can be contaminated with gluten during processing)
2. Bulgur	
3. Farina	8. Rye
4. Farro	9. Semolina
5. Freekeh	10. Spelt
6. Kamut	11. Triticale
	12. Wheat

---

[1] Aretaeus of Cappadocia is thought to have practiced medicine in both Alexandria and Rome. Besides celiac disease he has also been credited with describing the first cases of diabetes and asthma.

R̲X̲ **DNA *Restart Health Tip #4***
Here's a list of grains that do not contain gluten.

1. Amaranth
2. Buckwheat
3. Fonio
4. Millet
5. Oats (gluten-free certified)

6. Quinoa
7. Rice
8. Sorghum
9. Teff
10. Wild rice

In one of my previous books, *Survival of the Sickest*,[2] I described in great detail my theory that many common medical conditions are complicated "blessings." You may have inherited genes that predispose you to a condition such as high cholesterol, which offered your genetic ancestors some type of benefit or protection and allowed for their survival.

So are there any benefits to having celiac disease? Actually, a theory has been proposed that western Europeans are much more likely to get celiac disease because it made them anemic,[3] one of the complications of the condition. That may not make sense initially, but we now know definitively, with only rare exceptions, that all pathogenic bacteria are actually after your body's iron. This metal is the reason you get a fungal or bacterial infection in the first place. The more iron you have, the tastier a meal you would make in the eyes of pathogenic microbes (more on this later). This is why the better your body is at hiding its iron, the greater its ability to fight infections. I have spent the last 20 years investigating the biological and health implications of dietary metals, especially heavy metals like iron. Finally, after all those years, my research findings led me to discover a new antibiotic, the first in a new class of antibiotics that has been developed in more than 2 decades, that specifically targets "superbug" or resistant microbes like MRSA by interfering in the way they get and use iron. Since this discovery, I have stopped seeing patients and have been devoting my time and energy to developing powerful clinical interventions that have the potential to improve the quality of millions of people's health and life.

---

[2] If you're interested to know more about why and how many of the common diseases affecting people today actually helped your ancestors survive, read *Survival of the Sickest: A Medical Maverick Discovers Why We Need Disease.*

[3] Caused by a lack of iron.

So believe me, iron is a big deal: More than 50 million Europeans are thought to have died because of it due to the bubonic plague's many marches across Europe, beginning in the 14th century. Not everyone died, meaning some people survived. And it could very well be that the lack of iron, caused by celiac disease, would have helped play a small role to ensure the survival of some western Europeans.

THE FLIP SIDE: WHEN GLUTEN ISN'T THE PROBLEM

Celiac disease is not the only example of a condition where your body may have a problem with wheat. Sneezing, sniffles, and headaches can be symptoms of hay fever, but they can also be caused by a wheat allergy. This is an example of when the body overreacts to something in the environment that is usually not harmful, in this case wheat, and tries to fight it off as if it's a foreign invader.

It's not known exactly how common wheat allergies are, but they're thought to be less common than celiac disease in America. The immunological quarrelsomeness that causes wheat allergies is the result of the body making an IgE antibody against proteins that are found in wheat, just like what happens when you have a pollen, dust, or mite allergy, and this reaction then drives all of the symptoms of the condition.

If you think that you may be suffering from a wheat allergy, it's really important that you see an allergist to get tested. If you are diagnosed with an allergy to wheat and not celiac disease, then you shouldn't limit yourself or your family to a gluten-free dietary existence. Why not be able to eat rye and other whole cereal grains that do not contain wheat?

But what about the multitudes of people who do not have the classical diagnosis of either celiac disease or a wheat allergy but obviously have real symptoms associated with eating gluten? The one recurring thing I've often heard from patients, friends, and family is that many of them who have real issues with gluten are being labeled as having *non-celiac gluten sensitivity*.

That doesn't surprise me, given the many issues surrounding testing, especially the fact that you must be eating a lot of gluten weeks or even months prior to testing to test positive for celiac disease. Yet what was a small issue a few years ago has grown into an epidemic. And not everyone is happy with the label of non-celiac gluten sensitivity, or NCGS. Who wants to live with a label? People want actionable answers.

I've created a table to sort out some differences, including symptoms and causes, among celiac disease, wheat allergies, NCGS, and even irritable bowel syndrome (IBS).

Differences among Celiac Disease, Wheat Allergy, Non-Celiac Gluten Sensitivity, and Irritable Bowel Syndrome

Type of Condition	What Are the Symptoms?	How to Diagnose?	How Common?	What's the Cause?
Celiac Disease	Bloating Stomach cramps Diarrhea/constipation Iron-deficiency anemia Infertility, recurrent miscarriage Anxiety and depression	Blood test for immunoglobulin A tissue transglutaminase, or tTG-IgA test Intestinal biopsy	Most common in people with European ancestry Thought to affect about 1 percent of people worldwide	Autoimmune reaction to exposure to gluten-containing foods in sensitive people
Wheat Allergy	Skin rash/hives Stomach cramps Stuffy/runny nose Headaches Indigestion Anaphylaxis (rare)	Skin prick test Blood test for IgE antibody to wheat	Thought to affect 0.1 percent of people worldwide	Allergy to wheat protein
Non-Celiac Gluten Sensitiviy (NCGS)	Bloating Stomach cramps Foggy mind Aphthous ulcers (canker sores)	Excluding celiac disease and wheat allergy A double-blind gluten challenge	Not known	Not known
Irritable Bowel Syndrome (IBS)	Stomach cramps/pain Bloating Gas Nausea Diarrhea and/or constipation Frequent bowel movements	No direct diagnostic testing available	As many as 20 percent of Americans report symptoms of IBS	Not known

R
X
DNA Restart Health Tip #5

Here's a list of common food allergens to watch out for.

1. Cereal grains—such as wheat
2. Eggs
3. Fish
4. Legumes—such as peanuts and soybeans
5. Seafood—such as clams, crabs, lobster, and shrimp
6. Seeds—such as poppy and sesame
7. Tree nuts—almonds, cashews, and walnuts

Fortunately, there's been a lot more research trying to understand what's happening with people labeled with NCGS. There's now even a growing consensus that this large group of people is distinct from those who have celiac disease and wheat allergies. So if it's not celiac disease or a wheat allergy that's causing people so much dietary grief, what is the culprit behind our current dietary intolerance epidemic?

CHAPTER 5

The Dangers of Eating Soap

I want you to think for a minute about what you had to eat today. Did you have a bowl of yogurt? What about using a condiment like mustard?

It's very likely that at some point today you ingested a food product that contained an emulsifier. These are a group of chemicals that have detergent-like properties—yes, those are just fancy words for soap.

Emulsifiers are used within an enormous number of processed foods because they make them "shelf-stable" by helping to stabilize or keep ingredients from separating. In fact, it's hard to make your way through any of the middle aisles of your local supermarket without stepping on an emulsifier land mine.

So you might be wondering to yourself, what's the big deal with having a little emulsifier soap with your yogurt?

I started thinking that emulsifiers might be what's making us sick when I saw many patients who were plagued with chronic mouth sores called *aphthous stomatitis* or canker sores. When one of my patients, James, a high school principal in his sixties, forgot to pack toothpaste for himself and his son on their 7-day fly-fishing camping trip in Montana, something funny happened. All his canker sores cleared up.

That was until he returned home and started brushing with his usual toothpaste again. All his canker sores came back. James thought this might just be a coincidence, since canker sores have been associated with increased stress. Maybe he was just much more relaxed camping in Montana, he thought to himself. However, concerned that his conventional toothpaste could be the culprit, he headed to a local health food store to try a more "natural" brand, one that had a lot fewer ingredients for starters. He didn't particularly enjoy brushing with his new toothpaste, but just as he suspected, all of his canker sores completely resolved.

One ingredient that was missing from his more natural toothpaste was sodium lauryl sulfate, or SLS, part of a family of detergent-like compounds that help to foam up your toothpaste while you brush. And like all detergents, it's also good at cleaning things really well.

But for some people, a little too well.

And I happen to be one of those people. I was so surprised that James's change in toothpaste could solve his canker sore problem that I thought I'd give it a try as well. I am happy to say that since I switched to using toothpaste without sodium lauryl sulfate, I have never had a single canker sore. There have been some small studies looking at the effects of such ingredients in toothpaste, but the results have often been mixed, some finding a connection between the type of toothpaste you use and mouth sores, and others not. But as you're now learning by reading this book, what may be okay for some people can still be harmful for others.

It's not surprising that there have been such mixed results in these studies, as they often fail to take people's underlying genetics into account. Some people are obviously more susceptible to certain chemicals in their environment. I'll be speaking more about alcohol soon, in Chapter 9. Alcohol is an important example of the role that your genetics plays in your health; it's your genetics that determines whether or not you'll lower your risk of heart disease or conversely be at increased risk for esophageal cancer with every extra drink you have. That's why the DNA Restart takes your genetics into account, because it's so important to eat and live for your own unique genes and not someone else's.

But guess what? There's something else important to know about sodium lauryl sulfate: It can behave as an emulsifier. This means that by having detergent-like properties, it's also great at stripping away the protective layer from the surface of your mucous membranes. And to those who are susceptible, like James and me, this can cause painful canker sores. The more time I spent researching the connection between mouth sores and sodium lauryl sulfate, the more I became convinced that there were other possible human health consequences caused by other emulsifiers.

There's already unequivocal evidence that your gut's defensive barrier, which includes its mucous layer as well as the tight junctions between your gut's cells, is crucial to keeping inflammation in check. So what happens when we eat and drink foods that contain emulsifiers, and not just brush our teeth with them?

NO MORE EMULSIFIERS

Let's be very clear—we never genetically evolved the ability to eat the emulsifiers that have been added to the foods we are now consuming every single day. We simply do not have the DNA to eat detergents. Quite the contrary, in fact. So could it be that the consumption of emulsifiers that have been surreptitiously added to our foods in the last 50 years is devastating our gut, causing inflammation, and leading to leaks and trouble, just like they can when they're tucked in toothpaste?

I started thinking about this question a few years ago as another piece of the emulsifier puzzle fell into place. This had to do with the many cases I knew of individuals who were on a gluten-free diet in America, but "cheated" while traveling in

Europe. Samantha, a friend from college, told me that she couldn't resist sinking her teeth into a hot, freshly baked baguette while on her dream trip to Paris. Likewise, Jason, another physician colleague, told me about a similar experience not being able to resist having a Neapolitan pizza while traveling to Napoli, Italy. Samantha and Jason fell into that category I told you about previously of non-celiac gluten sensitivity, or NCGS, having tested negative for both celiac disease and wheat allergies, but apparently benefiting somewhat from a gluten-free diet.

When I asked for further dietary details from them, a pattern started to emerge: They both thought that they were going to pay rather dearly for their glutinous transgressions, as they always had in the past at home. Instead, something rather remarkable happened. Which was . . . absolutely nothing!

They were in complete disbelief as to how they could seemingly eat food that obviously contained gluten without getting ill. To say the least, so was I. That was, of course, until they returned to the United States and repeated eating gluten to disastrous effects. They both chalked up the experience in Europe to some type of lucky dietary break.

This got me thinking further. Could it be that what they experienced had nothing at all to do with luck? What came to my mind was all of the emulsifiers that have been packed into industrially prepared foods in recent years, such as commercially prepared baked goods like bread. This includes the artisanal, fancy, "freshly" baked bread at your local supermarket that's bursting with an enticing aroma, but may be anything but for your health. Unlike traditionally baked breads, these are stuffed full of emulsifiers such as dough conditioners.

These emulsifiers are added to help speed up the process of bread preparation and baking, which saves the supermarket both time and money. This allows bread to appear and smell fresh even though it was likely frozen and par-baked only a few minutes ago. And it's not just bread; hundreds of other processed foods now have added emulsifiers in them. We are literally drowning in a soapy sea of emulsifier soup.

Realizing this, I thought that it was time for a little experiment. What I did next was ask both Samantha and Jason to first go back to strictly following their previous gluten-free diet for the next 28 days. They both complied and let me know that they were feeling somewhat but not entirely better after the 4 weeks of being gluten-free. I then gave them an old family recipe for traditional slow-rise bread, similar to what they may have eaten in Europe, that's fully emulsifier-free, to try out at home. After some fun attempts at home baking, they both enjoyed a thin slice of the (emulsifier-free) bread and reported to me what they experienced. Which was . . . delicious bread that caused none of the previous symptoms that they had misattributed to gluten. It's been almost 2 years now, and both Samantha and Jason are still eating homemade bread—that's full of gluten, but free of emulsifiers—and feeling perfectly well. What could possibly be going on here?

I believe that the cause of the apparent rise in gluten sensitivity has nothing to do with gluten per se but everything to do with emulsifiers. The uptick in intestinal issues associated with gluten but negative for celiac disease and wheat allergies has been on a dramatic climb. Something has definitely changed in the last 50 years.

Interestingly, many people who have issues with gluten but don't have celiac disease or a wheat allergy also complain of aphthous stomatitis or canker sores—the same type of mouth sores I got when I used toothpaste with emulsifiers like SLS. Many commercially produced foods, including baked goods such as bread and cookies, now contain emulsifiers, which they never did in the past. This is why I believe that emulsifiers are to blame for many of our recent dietary problems associated with gluten. Eating and drinking foods that contain emulsifiers creates the perfect storm of leaky gut, inflammation, and a negative change in your microbiome that then makes us sensitive to food that previously kept us healthy.

The increase in consumption of foods containing added emulsifiers runs exactly parallel to the rates of increase in people reporting to be sensitive to gluten, as well as having irritable bowel syndrome (IBS). In fact, one in five, or more than 60 million, Americans report having symptoms consistent with IBS, which I believe for many of us is caused by unknowingly ingesting emulsifiers in our foods.

Just look back and follow the hidden emulsifier trail, and you'll see how it leads you right back to the multitude of food intolerances that have been associated with gluten. Within the DNA Restart, I've had shocking success with many people who have completely removed all emulsifiers from their diets and happily continued beyond the initial 28 days.

I have included a very comprehensive list of common and uncommon emulsifiers to watch out for and exclude completely from your and your family's DNA Restart (see page 35).

Eating out was a real nightmare. And even though I never got an official celiac diagnosis, cutting gluten from my diet made me feel somewhat better. It didn't make all my symptoms go away; there were some days when I felt just downright lousy. My problems with gluten are now all but gone! Now that I've gotten rid of all the emulsifiers in my diet, I can eat all of the gluten I want. I can't believe I suffered for so long. It's crazy, because I hadn't even heard of emulsifiers or known what they were doing to my body before becoming a DNA Restarter.

—Alison, 25

WHY YOU NEED TO MAKE YOUR LIFE EMULSIFIER-FREE

One of the most important things you will be doing in this pillar, Eat for Your Genes, is for the next 28 days on the DNA Restart—regardless of your current relationship with gluten—you will follow the complete emulsifier food exclusion rule. *And there are absolutely no exceptions here.*

What I want you to do next may be a little painful, but it's totally worth it. You will need to start the purging process at home, just as both Samantha and Jason did—that way, you'll remove any slight chance of an "accidental" binge driven by the late-night munchies.

DNA Restarters have used this same list to completely purge their households of all beverages and foods that contain emulsifiers. They also took the list with them whenever they went food shopping. Keep in mind that it's not a good idea to go on a gluten binge if you've been gluten-free for years, even if you suspect that emulsifiers are the real culprits for your digestive woes. That's because it may take 8 weeks or longer for your digestive system to heal from a lifetime of eating emulsifiers.

Given the ubiquity of emulsifiers in our prepared food chain and our chronic daily exposure, it's no wonder that we're seeing epidemic proportions of intestinal troubles. A bigger problem is that many of the commercially prepared foods that are gluten-free are not always emulsifier-free. Some of them actually contain even more emulsifiers than their gluten-filled doppelgängers! So you need to be very attentive and refer to your DNA Restart Emulsifier-Busting List to stay ahead of food manufacturers who are constantly trying to change and hide the names of the emulsifiers they often use, including the codes and acronyms, in the products they create.

At the end of the day, we should all be eating emulsifier-free. If you're the type who needs a little scientific motivation to start kicking emulsifiers out of your life, there's now research that can help you with that.

EMULSIFIERS ARE MAKING US SICK

In a sophisticated study published in the journal *Nature*, scientists at Emory University decided to test two common emulsifiers that are found in many commercially produced foods. They wanted to see what would happen if you fed mice relatively low concentrations of carboxymethylcellulose (CMC) and polysorbate 80 (P80).

What they saw was totally unexpected, since both of these common emulsifiers have a Generally Recognized as Safe (GRAS) designation from the FDA. The researchers noted a low-grade inflammation in the intestines of the exposed mice, who are usually otherwise very healthy. The study was rather elegant, as they not only tested these emulsifiers in regular mice, but they also decided to feed the emulsifiers to mice that were genetically prone to having gut inflammation. By doing this, they were mimicking what would happen if someone just like James or maybe even you were to ingest these emulsifiers—that is, someone whose genes make them susceptible to inflammation. Both groups of mice seemed to have a breakdown in the mucosa that's normally found in the gut. But the mice that were prone to inflammation fared even worse than their healthier cousins. Sadly, some mice even went on to develop full-blown colitis.

It's this change in the intestinal landscape that caused another rather surprising thing to happen. It brought the microbes that normally inhabit the gut precipitously much closer to the cells that normally line the gut. Imagine how you would feel if you were standing in a line and a stranger started to edge his way ever closer to you, to the point that you started to feel his hot breath on your neck. Would that make you happy? Probably not.

Now imagine what those poor gut cells felt like, having to be so close to bacteria that they are usually safely far away from. This uncomfortable distance then seems to have triggered even more inflammation. No wonder the mice got sick! You'd think that should be enough to get you to never take another bite into foods that contain emulsifiers—but wait, because there's even more.

EMULSIFIERS ARE MAKING US FAT: INFLAMMATION, LEAKY GUTS, AND OBESITY

It wasn't just the thinning of the layers of mucus that act as a safety barrier in the intestines of mice that alarmed the researchers in the emulsifier study. Eating and drinking food that had emulsifiers was triggering something even more surprising in the mice. The detergent-like properties of the emulsifiers also seemed to make their guts leakier.

Having increased intestinal permeability, aka leaky gut, and lots of microbes around is never a good idea. Like a garbage bag full of holes, a leaky gut will make a mess that is difficult to clean up. And that mess includes intestinal contents that are now able to get into your bloodstream. Yes, that's just as disgusting as it sounds.

Now if you needed just one more reason to eliminate all of the added emulsifiers from your life, there's just one more finding from that study that I'd like to share with you.

And it's this: The mice got fat!

The emulsifiers not only threw their glucose control out the window—think metabolic syndrome—but some of the mice also got both really hungry and then, upon frantic overeating, really fat. The only mice that didn't get fat were the ones who got full-blown colitis—maybe they were just too sick to eat. So what could possibly be making the mice that were fed emulsifiers so hungry?

HOW YOUR GUT MICROBIOME CAN MAKE YOU FAT, OR SKINNY

If you're not familiar with your gut microbiome, don't worry, because the microbes that constitute it are already very familiar with you. From the day you start your life, you begin collecting an array of microorganisms that make your body their home. And your diet in particular can affect which types of microbes inhabit your gut.

It's not just the gut that has its own unique microbiome. Every nook and crevice on your body is lined with its own unique ecosystem of microbes—imagine the differences among desert, mountain, and rain forest. Most of your microbiome helps to keep you alive and healthy by keeping out all of the unwanted intruders who are out to make your life miserable. As I've mentioned earlier, many of the nastier microbes would want nothing more than to clear-cut your friendly microbial forests and strip-mine all of your iron.

Sometimes changing your body's environment can have surprising consequences. Few people could have imagined that the rise in popularity of full Brazilian waxes would change the pubic landscape so much that lice that used to inhabit it have now become homeless. And that's a good thing.[1] But the change in the microbiome that happens in the gut because of the consumption of emulsifiers could never have been predicted. And the latest research now confirms that it's actually the microbiome that plays a big role when it comes to inflammation and obesity.

How do we know this?

Many previous experiments have shown that a type of bacterial family called Bilophila, for example, is associated with inflammation of the gut and promoting obesity. In the study described above where mice were exposed to emulsifiers, the researchers also saw a decrease in health-promoting Bacteroidales and an increase in inflammation-promoting Proteobacteria. To see what the changes to the microbiome could be causing when the mice were given emulsifiers, the researchers went one step further.

They transferred the microbes from mice that were chronically exposed to daily low doses of emulsifiers to an unexposed group of mice, who then experienced a big change in their microbiome. And when they transferred the emulsifier

---

[1] In one of my previous books, *How Sex Works: Why We Look, Smell, Taste, Feel, and Act the Way We Do,* I discuss how many of the changes that have occurred in modern times have impacted our sexuality.

microbiome to other mice that were on an emulsifier-free diet, those mice also became *sick* and *fat*.

What further really concerns me regarding this study is that the mice became sick from the emulsifiers at very low concentrations. In fact, the amount of emulsifiers that was used in one study was 20 times less than some people are exposed to every day! There still remains a lot more research to be done to understand the extent to which emulsifiers harm your health.

With more than 60 million Americans reporting symptoms of IBS, my hope is that the FDA will take action and revise the Generally Recognized as Safe (GRAS) status of these dangerous emulsifiers that we are unknowingly eating every day. But that doesn't mean that you need to wait for more research or a governmental agency to change their policy. You can start taking your health into your own hands and take action now. Why wait?

It's time for you to head over to your kitchen and start purging your life of unnecessary emulsifiers (refer to the DNA Restart Emulsifier-Busting List for a complete listing).

For those people who have been dealing with digestive issues, or who have been diagnosed with IBS, I don't want you to feel discouraged if you have any intestinal symptoms that don't improve right away after you've stopped all of your exposure to emulsifiers. From the research that has been done, it seems that it can take about 8 weeks or even longer for a reversal of some of the inflammation after a chronic exposure to dietary emulsifiers. And if you're up for trying to reintroduce gluten after your 28-day DNA Restart, remember my advice to Samantha and Jason to go slow.

Why not ditch the emulsifiers and start living your life the DNA Restart way? The science suggests that you will actually start eating less, lose weight, stop inflammation, normalize your glucose, and stop having a leaky gut. I don't know about you, but that sure sounds like a great way to spend the next 28 days. If you are still having any digestive issues that include bloating, cramping, or diarrhea—which can sound like IBS— after the 28 days, then I want you to go back to the DNA Restart Emulsifier-Busting List and scrutinize the ingredients in your food, looking for hidden emulsifiers.

If you're one of those lucky people who have never experienced any ongoing digestive issues, I still think it's important and worthwhile for you to go emulsifier-free. You might be experiencing a chronic low-grade intestinal inflammation that is triggering you to eat more, get fat, and lose control of your blood glucose levels without even knowing it. With all of the trouble they cause, are emulsifiers really worth eating?

The DNA Restart Emulsifier–Busting List

Codes Used	Common Emulsifiers	Some Common Food Products That Contain Them
E442	Ammonium phosphatides	Animal products, dairy products
E482	Calcium stearoyl-2-lactylate	Baked goods
E466	Carboxymethylcellulose (CMC)	Baked goods, egg whites, dehydrated potatoes
E407	Carrageenan	Dairy products, baked goods
E472e	Diacetyl tartaric acid	Baked goods, salad dressings
E488	Ethoxylated monoglyceride	Baked goods, dairy products
E412	Guar gum	Dairy products
E322	Lecithin	Baked goods, chocolate
E410	Locust bean gum	Baked goods, dairy products, liquors
E435	Polysorbate 60 (P60)	Baked goods, salad dressings, dairy products
E436	Polysorbate 65 (P65)	Baked goods, dairy products
E433	Polysorbate 80 (P80)	Baked goods, dairy products, frying oil
E477	Propylene glycol monostearate	Baked goods and dehydrated potatoes
E481	Sodium stearoyl-2-lactylate	Baked goods, dairy products
E491	Sorbitan monostearate	Dairy products, cocoa
E420	Sorbitol	Baked goods, sugar substitute
E472g	Succinylated monoglyceride	Baked goods
E473	Sucrose monostearate	Baked goods, dairy products

CHAPTER 6

Eat Dairy If Your DNA Lets You

Now that we've covered your optimal carbohydrate intake with the DNA Restart Cracker Self-Test and committed to banishing all emulsifiers from your diet, it's the perfect time to move on to talk about dairy.

No one knows for sure who was the first person to decide to reach down and grab a ruminant animal's teat and have a drink, but I think we should call him Pavel. Yes, I know what you're thinking . . . Pavel was some kind of a mutant. And, actually, from a genetic perspective that's true! That's because if Pavel didn't get bloated and have gas or diarrhea, then in genetic terms we would say that he's *lactase persistent*, or LP. He didn't get sick after having milk as an adult because he had a mutation in his DNA that allowed his *LCT* gene to make the enzyme lactase, to break down lactose (the sugar in milk).

Every mammal, except for some humans today, turns off their *LCT* gene after they complete weaning as babies. The reason, of course, is that mature adults usually don't drink milk intended for babies. This helps explain why most people on Earth today cannot have milk as adults without getting sick—they cannot digest the sugar lactose that's found in milk. Lactose, in someone who's lactose intolerant, then finds its way into the colon, where it's gladly digested by the microbiome in the gut. And that's where the gas and bloating come from—it's actually the bacteria inside the colon producing hydrogen gas from their nutritious lactose meal, which then leaves you passing wind or, simply put, making farts. So don't let the milk board ads fool you. Milk doesn't do the body good for most people alive today. But as you know by now, the DNA Restart is not about other people. It's about you.

So that's the dairy reality for a shocking two-thirds of the world's adult population. So why are some people genetic mutants who are able to have lactose while most others cannot? It all depends on whether or not you had ancestors who were agriculturalists and kept animals for their milk. Some of these ancestors benefited not only from the simple life-sustaining calories but also from the extra nutrients, fats, and carbohydrates they acquired from milk and products made from it. This, of course, is especially true when, as every good farmer knows, plant-derived foods become scarce between harvest times, such as the middle of winter.

Our genes have an immense ability to determine what we can and cannot eat. If you inherit certain DNA, you can enjoy a banana split sundae without the accompanying bloating and stomach upset encountered by someone who was born to be lactose intolerant. As I'll explain in much greater detail later, our ability to eat dairy products as adults is a great example of the power of our genes to determine what we can or cannot eat. If you're born with an *LCT* gene that works when you're an adult, that's great, especially if you love foods with lactose in them, because the *LCT* gene codes for an enzyme called lactase that breaks down dairy sugar.

However, it's not just the particular version of a gene that you have inherited from your parents that determines if ice cream leaves you feeling unwell. It also happens to be that the number of *AMY1* genes you've inherited determines how many carbs you can eat. Genetics is like that—the more you start digging, the more meaningful genetic differences you discover between people.

If you're one of those people who have no problem consuming milk, then you're in the lucky minority on this planet. These lucky few have inherited one of only a handful of known mutations. And at some point in your ancestors' lives it would have meant the difference between life and death: When there's nothing but milk to drink, it can sure do the body good. Pavel's mutation kept his *LCT* gene turned on well into adulthood, pumping out all of the lactase needed to digest the lactose in the milk he was drinking.

Well, that milk-loving mutant I was telling you about was an actual person (although his name was likely not Pavel) who lived sometime around 8,000 to 12,000 years ago in what is today eastern Europe. What's fascinating is that every person alive today who has European ancestry and is able to have milk shares Pavel as an ancestor.

This genetic ability to enjoy dairy as an adult has spread quite nicely around the world, but every time it happened in a group of people, it was a brand-new genetic mutation. We know of only a handful of mutations that let people drink milk as adults because they all have lactase persistence, or LP. And where your ancestors come from will determine if you can drink milk today.

So what's your dairy status?

You already likely know if you are lactose intolerant or not. If you experience bloating, abdominal cramping (especially around your belly button), diarrhea, flatulence, and a loud rumbling in your stomach after eating foods that contain lactose, it's very likely that you are lactose intolerant. And if that's the case, I want you to continue not having any foods that contain lactose for the next 28 days.

In contrast, if you are able to have dairy products without any issues, there's no need to exclude these foods. There's a reason your genetic ancestors gave you the genetic know-how to break down lactose as an adult. When dairy products are fermented (such as kefir and yogurt), they are an especially potent source of pre- and probiotics. So you'd be smart to continue to eat the way your genetic ancestors did for thousands of years.

WHEN DAIRY'S NOT THE PROBLEM: EMULSIFIERS AGAIN!

It's not always so simple to know if a digestive issue is directly related to the consumption of dairy that contains lactose. There are other reasons why dairy might cause digestive problems, and they can include an allergy to milk protein, which is similar to having a wheat allergy. Or it can be the result of an autoimmune disease that involves both the small and large intestines, such as Crohn's disease. There can also be problems with dairy when the microbiome in your gut is out of balance, such as when there's bacterial overgrowth or, less commonly, in situations of infections.

There's also been a growing trend of people experiencing problems with dairy, not unlike what we've spoken about regarding gluten. This is something I actually experienced myself a few years ago. As a teenager, I started experiencing severe abdominal pains whenever I had certain dairy products. Now, it sounds like what I was experiencing was a very common case of lactose intolerance, which can happen when your body doesn't make enough of the enzyme lactase that breaks down and digests the sugar lactose that is found in milk.

What was a little confusing was that there were times when I seemed to be completely unaffected and could easily and happily consume ice cream that was chock-full of lactose. But then out of nowhere, after just a small teaspoon of sour cream, I'd get sick again. I could even drink a glass of regular full-fat milk, but sometimes when I tried to drink some brands of chocolate milk, I'd find myself doubled over in pain. So this didn't sound like a problem of my body not making enough of the enzyme lactase to break down lactose, as the glasses of whole milk I was drinking were full of that sugar. My family doctor at the time thought I might have IBS.

As a young budding scientist, I was unhappy with my doctor's uncertainty at the time. "Might" just didn't cut it for me. So I went ahead and turned my dietary mystery into a high school science project.

To create an accurate inventory of my symptoms, I decided to keep a food diary for an entire month. When the month was over, I pored over my data but couldn't find any significant patterns—except that Sunday afternoons seemed to be a time I would commonly suffer from the "milk-runs," as I started to affectionately refer to them.

And then it hit me.

Sunday afternoons were when my family typically would have a brunch with friends and family, a great time where everyone brought dishes of food, which we shared and ate communally. This was at the height of the "fat-free" dietary fad that afflicted much of America at the time. And so many of the dishes that our friends brought over reflected that. The dairy products seemed to be hardest hit, with labels proudly advertising their lack of a drop of fat.

My parents, on the other hand, embraced more traditional products that were less adulterated and had the fewest ingredients. That usually meant that my family ate foods and ingredients that were full-fat products, prepared through more traditional means, unlike the rest of the scientifically crafted, laboratory-inspired foodstuffs at the table. Because of this, the only time I would ever eat any fat-free foods was unintentionally at a restaurant, or at my family's weekly communal brunch.

Combing over my notes, I verified that the only times I'd get sick were either if I ate dairy products out or if I would eat a particular dish one of my parents' friends would bring to our weekly brunch. So I began to think that the culprit must be some kind of additive in the fat-free dairy products.

To test out my hypothesis, I asked my parents if they would mind if we started going over to their friends' houses for dinner in addition to the weekly brunches. Lo and behold, the milk-runs revisited midweek with a vengeance. The almost instantaneous abdominal cramping would only happen after eating dinner at some households and not others.

My next move was to ask to see all the ingredients in the food that I ate, as I became convinced that my symptoms had to be due to something I was eating only from a few particular kitchens. But with so many ingredients, I couldn't put my finger on what was making me so sick. The final breakthrough came one hot day when I stopped by a convenience store on the way home from school and bought a chocolate milk in a desperate attempt to quench my thirst. Getting back on my bike and starting to pedal, I realized something was terribly wrong. I mean really, really wrong. The milk-runs were back. What was in this killer chocolate milk? The next day I returned to the same store to investigate what was in the drink I bought the day before. These were the ingredients that were listed: *Skim Milk, High Fructose Corn Syrup, Cocoa processed with Alkali, Corn Starch, Salt, Artificial Flavor, Carrageenan, Vitamin A Palmitate, and Vitamin D$_3$.* As I looked closely at every single ingredient, I noticed there was something listed that I didn't recognize. Could that be what was making me so sick?

Carrageenans are a "natural" group of chemical compounds that are extracted from red seaweed and have been indispensable to food manufacturers and corporations looking to increase their profit margins. Citizens of countries belonging to the European Union are a little more shielded from carrageenan because these countries adhere to tighter restrictions as to which products can contain them; for example, infant formulas are free from this additive. But in the United States, carrageenan has made its way through the processed food chain and is included in everything from shoe polish to vegetarian hot dogs, chocolate milk, toothpaste, and many types of dairy products, such as ice cream and yogurt. It's disturbing when you realize that emulsifiers have even made their way into some organic infant formula. So please be diligent and refer to your DNA Restart Emulsifier-Busting List to avoid many common products that contain carrageenan and other potentially dangerous emulsifiers.

The reason carrageenan has become so ubiquitous in our processed food chain is its special ability to act as a thickening agent—I like to think of it as a cheap food glue. And this works great when you want to make products that are unnaturally low-fat or fat-free or if you want to make a cheaper dairy product using powdered milk and increase your profits that way.

There's also a long history in Asia and Ireland, in particular, of consuming a mildly processed form of red algae that naturally contains carrageenan. But that doesn't mean that it's safe for everyone. Remember what you now know about your genes and diet—we should only be eating for our genes and for *our* genes alone. Some people may just be genetically predisposed to carrageenan sensitivity.

So why was carrageenan making me sick?

It turns out that carrageenan has a rather colorful laboratory history. For one, back in the late 1960s, scientists discovered that it is possible to get a model for inflammatory bowel disease by feeding guinea pigs . . . let's see if you can guess? That's right, carrageenan! Once the poor guinea pigs got enough of carrageenan, their guts became irreversibly harmed. Does this sound familiar? As it turns out, if you haven't guessed already, carrageenan is also an emulsifier. Don't let anyone ever convince you that it's totally safe for all humans to eat it. Thankfully, I discovered early in life that it wasn't dairy that was bothering me, but rather a cost-saving food ingredient.

The reason I think that carrageenan irritates the gut for some people is that it can resemble a nasty microbial biofilm. This is the gooey-sticky gel or slime that microbes can produce to give them a nice, protected home. Think of the green biofilm or slime that can cover the glass of a fish tank and you'll get the picture. Carrageenan can resemble these biofilms, alerting your body to a possible microbial invasion.

Then there's the DNA you inherit that can predispose you to a worse reaction to all emulsifiers. This is in line with the emulsifier-mouse study I told you about previously. Remember, some mice got sicker than others when they were given emulsifiers to eat.

Genetic predispositions aside, I don't believe emulsifiers in any quantity are good for you or your genes. If you have any digestive issues that include bloating, cramping, or diarrhea—which can sound like lactose intolerance, NCGS, or IBS—I want you to go back once again to your DNA Restart Emulsifier-Busting List and scrutinize the ingredients in the dairy products you're eating to look for hidden emulsifiers.

If you still want to try including dairy products but think you might actually be lactose intolerant, I've also included a list of some common lactose-containing foods and the corresponding amounts of lactose in the chart below. If you think that you may be a little sensitive to lactose, you should be consuming only those foods in the chart that contain the lowest amounts of lactose.

Some types of hard cheeses (the harder the variety, the better) contain less lactose and can be eaten even by those who are generally lactose intolerant. You can start by trying foods that have the least amount of lactose and that also have a solid consistency. Hard cheeses like Parmesan and aged Cheddar, which naturally contain the least amount of lactose, would be a good place to start in small quantities.

Your Lactose Guide for Some Common Foods

Food	Amount of Lactose (Grams)
Parmesan (1 oz)	0.05
Cheddar (1 oz)	0.5
Kefir (1 cup)	0.5–3
Mozzarella (1 oz)	1
Whey protein powder (100%) (30 g)	1–2
Cottage cheese (½ cup)	3
Lactaid milk (1 cup)	3
Ricotta (1 oz)	3
Yogurt (1 cup)	4–18
Greek yogurt (1 cup)	9
Goat milk (1 cup)	10
Milk: whole, low-fat, fat-free (1 cup)	12

Another strategy is to do what your genetic ancestors might have done: Consume only those milk products that have undergone fermentation (such as kefir) and therefore naturally have less lactose present while being full of pre- and probiotics.

CHAPTER 7

When People Pump Too Much Iron

Most of us are getting too much iron from our diets today.[1] But wait a minute; isn't getting a lot of iron a good thing? The answer is not so simple, as you may remember from Thomas's story in the beginning of this pillar. The reason has to do with the fact that even though we all need some iron for optimal health, once you start getting too much, you can quickly run into some serious health problems. And our genes play an incredibly powerful role in deciding just how much iron we need.

For example, *hereditary hemochromatosis* is a genetic condition that is caused by specific changes in the DNA of the *HFE* gene that you inherit from your parents. If you inherit hemochromatosis, your body is like an incredible giant magnet working to pull out and absorb as much iron as possible from what you're eating every day.

Although we all need iron to serve as the engine driving much of the cellular machinery that makes up our bodies and to make enough life-sustaining hemoglobin, how much iron your body actually needs is determined fundamentally by the genes you inherit. Get the DNA for hereditary hemochromatosis, and you'll also find yourself slowly but surely rusting to death on a diet that might lead to longevity for someone else.

At least one million Americans are thought to have hemochromatosis today, but most of them are actually completely unaware that they have the condition, and yet are slowly rusting to death while enjoying a nice steak dinner. To make matters worse, current nutritional guidelines for iron do not take any genetic variability into account and, if followed to a T, can seriously harm people with hemochromatosis.

For example, to help make sure that everyone gets enough iron—especially women, who can be deficient because they lose iron when they menstruate—it is even added to flour. That's what it means when the label reads "fortified" or "enriched."

---

[1] With the exception of some menstruating and pregnant women who have higher iron needs and can develop anemia due to a higher demand on their stores of iron.

But today, as we eat not only more fortified foods but also more meat, our iron intake has increased significantly. In fact, it's thought that on average our intake of iron has doubled in the last 50 years. And when it comes to iron, eating too much means you're asking for trouble. That's because excess iron can wreak havoc in the body, causing everything from increased inflammation to fatigue, diabetes, joint and muscle pain, erectile dysfunction, arthritis, and even cirrhosis of the liver.

In the case of someone with hemochromatosis, as organs such as their heart, pancreas, and liver start filling with extra iron, they also start to experience increased levels of oxidative stress. And that's why their bodies start to fail. They are literally rusting from the inside out.

If you are fortunate enough to get diagnosed early, the treatment consists of making dietary changes to lower your iron intake and having routine phlebotomy, or bloodlettings.

To find out if you're at risk of rusting to death, you can ask your doctor to add three simple blood tests to your next annual health checkup. The risk is highest for men and postmenopausal women because they don't menstruate or experience pregnancy, two ways in which iron levels are naturally reduced.

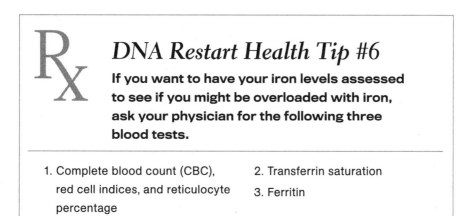

DNA Restart Health Tip #6

If you want to have your iron levels assessed to see if you might be overloaded with iron, ask your physician for the following three blood tests.

1. Complete blood count (CBC), red cell indices, and reticulocyte percentage

2. Transferrin saturation

3. Ferritin

CHAPTER 8

Limiting Red Meat and Banning Processed Meats from Your DNA Restart

According to the beef industry, red meat has a lot of positive things going for it. It's a great dietary source of both protein and essential minerals like iron. There's only one small wrinkle when it comes to red meat: Like all good things, it's very easy to get too much of it.

Eating too many vegetables, of course, can also be problematic. Thanks to the beta-carotene found in carrots, some people have turned themselves orange by juicing and consuming way too many of them. Turning orange is one thing, but developing cancer and dying young is quite another. That's because too much red meat kills—it's as simple as that. Well, maybe not right away like, say, ingesting a cyanide pill, but more like arsenic in your drinking water or food. The more you consume, the faster you're guaranteed to die.

It's not just people with hemochromatosis who are getting into trouble by eating too much red meat. Just like with the current overconsumption of processed sugars, more than likely your genetic ancestors were never in a position to eat the amount of red meat you're consuming today. That's why as part of the DNA Restart, you are going to limit your red meat intake to only two servings per week. You're also going to completely cut out all processed meats. This includes bacon, beef jerky, corned beef, ham, hot dogs, pastrami, salami, and sausages as well as canned meat and any other lunchmeats. If you're willing to go further and cut out meat completely from your diet, I'm not going to stand in your way. We must face up to the fact that unlike other carnivores—mammals that eat only meat—it's clear from our anatomy and physiology that we can easily do without red meat, especially if we eat poultry and fish (and I'll be explaining to you later why I far prefer you to indulge in the latter). I realize you might not want to stop eating red meat because you think that making this change is going to be difficult. And it may very well be. You may get a lot of resistance from your red-meat-loving friends. But I can assure you that reducing your red meat consumption is not going to be as difficult as, say, having to deal with breast or colorectal cancer.

In fact, it's not red meat that we can't live without but actually fresh fruits and vegetables. (I'll be speaking about this in more detail specifically in the 2nd Pillar: Reverse Aging.)

The more scientists probe deeply into the health effects of red meat consumption, the more negative health problems they seem to find. Topping the list currently, according to the American Institute for Cancer Research and the World Cancer Research Fund, is an increased risk for colorectal cancer. There's also another laundry list of other cancers that have been suggested to be linked to the consumption of red meat. And, of course, it's not just cancers that should worry you, but also cardiovascular issues, diabetes, and hypertension.

There have been a few reasons proposed for the ill effects that eating red meat has on your body. One of the reasons has to do with the change in your microbiome that happens from eating a lot of red meat. One of the goals of the DNA Restart is to make sure that you get enough dietary fiber, which helps favor the growth of beneficial microbes, which can help shift your microbiome in a positive direction. Many of these beneficial microbes help dampen and reduce inflammation in the gut, as well as produce butyrate, which is used as a food source by the cells that line your intestine.

When you're eating a lot of red meat, your microbiome is thought to be responsible for converting carnitine—which is naturally present in meat—into a toxic compound called trimethylamine-N-oxide, or TMAO, which wreaks havoc on your cardiovascular system. This microbial conversion of carnitine seems to happen more abundantly in people who eat more red meat, making it another good reason to stop

R$_X$ DNA Restart Health Tip #7

Consume only two servings of red meat a week and absolutely no processed meats.

Although a great source of protein, red meat can increase your risk of getting cancer and heart disease. That's why on your DNA Restart you will need to limit your intake of red meat to two portions (2 to 3 ounces each) a week. Additionally, no processed meats are allowed during your DNA Restart.

eating loads of red meat. There are also other chemicals that get produced when you ingest meat. When you mix lots of protein with iron, for example, you get a group of not-so-happy chemicals called N-nitroso compounds, or NOCs for short. These are really potent carcinogens for the gut. And that's no good for anyone.

Then, of course, there's the problem of antibiotic use in the production of meat. As our meat consumption has increased considerably over the last 50 years, so has the amount of antibiotics used in its production. The price has also dropped precipitously, thanks in large part to how meat is produced today. It's essentially been industrialized—that picture you see on your shrink-wrapped package of ribs of a red barn and a single cow being lovingly stroked by a farmer is, for all intents and purposes, a complete and utter fabrication.

Most of the cows and pigs that people are eating today have likely never been outside the building that warehouses them. And the closest they ever get to standing outside a red barn with a farmer chewing on a stem of wheat is in a fever-induced delirium. That's because besides making the raising of beef and pork more profitable, the industrialization of meat production puts quite the physiological strain on the animals we ultimately eat. This means they get sick quite easily. The results are often infectious diseases and animals that are completely unfit for your dinner table. This also results in a loss for the meat producer.

This is one of the reasons why the meat industry started to use antibiotics. The other reason for giving animals antibiotics is that it increases their rate of growth, which ultimately increases profits because it results in bigger animals even though they're eating the same amount of food.

Yet these benefits come at what cost to our health and well-being?

Not only have Americans today done themselves a dietary disservice by eating larger amounts of red meat, but they've also unknowingly played a major role in creating "superbug" or multiresistant microbes. I should know, as I've spent decades researching new ways to deal with superbug infections. Realizing as a physician and scientist that our antibiotic arsenal is sorely lacking to fight these new superbugs, I have sharply focused my research on finding new types of antibiotics that can deal with the current and emerging threats.

Yet all of this work will be in vain if meat producers are allowed to feed the antibiotics I've developed to their animals just so they can make a profit at all of our expense. Thankfully, there is talk of limiting the ability of meat producers to use antibiotics, but for the most part it's too little too late, as many killer microbes have already been bred from past inaction. So do as the DNA Restart demands and limit your red meat consumption to no more than two servings per week.

CHAPTER 9

The DNA Restart Optimized Alcohol Intake Guide

Before you uncork that bottle of bubbly to start celebrating the beginning of your new genetic life, let's get one important question out of the way. How much alcohol is good for you and your genes? To answer this question, I've developed a genetic self-test so that you can discover your unique, genetically optimized level of alcohol intake. And to do that, you're going to have to get really intimate with a part of your body you may not have given much attention to up until now. The results of this next self-test have the power to substantially lower your risk of developing the deadliest cancer known to medicine.

It's time for us to scientifically investigate the genetics behind your earwax and how you can use that information to optimize your health.

DRINK ALCOHOL ACCORDING TO YOUR EARWAX TYPE

By now I'm sure you will not be surprised when I tell you that some of us are more genetically endowed with an ability to break down alcohol than others. You may have already discovered this human genetic quirk when you have had a drink or two (or maybe three or more) with friends who come from different genetic ancestral backgrounds. So what you already know is that not all drinkers are created equal.

What you may not know is that the key difference when it comes to how your body handles alcohol has to do with the genes you've inherited. And these genes can have an immense impact far beyond just predisposing you to a quick and nasty hangover. They can leave you with a devastating cancer.

Now I don't just want you to be eating for your genes, but drinking for them as well. So I've devised a genetic self-test that helps to very quickly assess which part of the world your genetic ancestors came from. This is very useful today, especially in the United States, as we all for the most part descended from some type of mixed ancestry.

As I just mentioned, the genes you've inherited from your ancestors greatly determine your ability to break down alcohol. For the most part it's a two-step process, as you can see in the figure below. Alcohol in the form of ethanol (regardless of which type of beverage you're drinking, it's all ethanol at the end of the day) is converted first into *acetaldehyde*, and finally into *acetic acid*.

The Steps of Alcohol Detoxification in Your Body

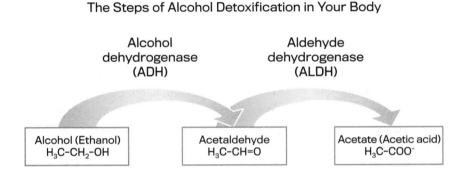

So why should you care about the complex genetics behind the chemistry of alcohol detoxification? Because, depending upon the DNA you've inherited, your next glass of red wine, or any other alcoholic beverage for that matter, can either be good for your heart or set you up for one of the deadliest cancers in the world.

Squamous cell esophageal cancer is the result of DNA changes to cells that line the mouth and the throat. Survival rates for this type of cancer are not good—only 15 percent of people are still alive after only 5 years.

These cells that line the mouth and the esophagus turn cancerous because their genes were damaged not by the alcohol per se, but by acetaldehyde that builds up in people who lack the genetics to easily detoxify this poison. It's not just in the throat that acetaldehyde can build up and cause mutations in your DNA. If you have inherited certain genes from your ancestors, every sip of your martini causes your body to convert the alcohol in your drink into thousands of knives that then shred your DNA. This shredding causes deadly mutations that can lead to cancer.

So why do organizations like the American Heart Association (AHA) recommend that both men and women consume alcohol every day? To be honest, it's actually a great recommendation with a lot of good science to back it up—limited intake of alcohol is evidently good for your heart. There's only one serious problem with such proclamations. The AHA is committing the same mistake as your doctor, and everyone else who has been giving you health and dietary advice over the years.

They never take your specific DNA into account. This is another example of a good intention gone bad. Because when it comes to your DNA and the way it behaves, you are not like anyone else. And because we're speaking about the way in which your body processes alcohol, the most common difference has to do with the *ALDH2* gene. If you happen to be born with just one gene that's marked *ALDH2*2*, for example, then you're like the hundreds of millions of other people on the planet who are at a much higher risk of developing squamous cell esophageal cancer the more alcohol you consume.

There's one easy way to know if you or someone you know is at risk from drinking. If your face turns red when you're drinking, you're experiencing a histamine release because of a buildup of acetaldehyde.

This method is not foolproof, as not everyone flushes in this way every time they drink, and other genes are involved that can put you at risk. But wouldn't you like to know if you're starting to shred your DNA when you're moving on to your second or third drink?

This is why I'm going to help you find out your optimal alcohol intake level by using an indirect route, which I've developed specifically for the DNA Restart. But to do that, we're going to have to talk about something you probably don't think about very much—your earwax.

Humans are born with one of two types of earwax, or *cerumen*. Earwax may not be the sexiest thing for you to examine carefully, but it's going to be an important clue to understanding your genetic ancestry. Earwax is actually produced by glands in the canal of the ear.

If your earwax is of the *wet* type, you likely have European or African ancestry. Wet-type earwax usually has a yellow or brownish color and is rather sticky. However, if your earwax is of the *dry* type, more likely than not you have ancestors who

---

It was a bad joke in my family because as kids growing up, both my sister and I could never hide from our parents whenever we went drinking, because we turned beet red. In college I kept having to explain to my fraternity brothers that Asian flushing didn't mean that I couldn't drink, it just meant I turned embarrassingly red in the face. I figured out that if I took an antihistamine before a drinking session, I wouldn't turn as red. But until I went on the DNA Restart, I had no idea that my "flushing" from drinking was a sign that I was at increased risk for cancer with every beer. All that from a Q-tip! I have totally cut down my intake now.

—Peter, 42

originally lived in eastern Asia. The dry type of earwax can be described as usually flaky and a little lighter in color.

In genetics we refer to these binary wet/dry earwax types as a binary trait. It was only in 2006 that the gene *ABCC11* was found to be responsible for the type of earwax you ultimately inherit. So by looking at your earwax, and determining which type you've inherited, you can also find out something about your genetic ancestry. This is going to be the key to unlocking a more optimal alcohol intake level for you.

So let's get started!

For this DNA Restart Self-Test you are going to determine the amount of alcohol that you are allowed to consume every week. To find out your optimal alcohol level intake, you are going to need a few cotton swabs and a piece of paper.

I'd like you now to get your cotton swabs and carefully clean out each ear[1] (use one swab per ear) and then place each of the swabs on a single sheet of plain white paper. Now tap or wipe each end of the soiled swab on the piece of paper.

Now have a closer look at your earwax. What color is your earwax? Is it moist and sticky? Or does it appear dry and flaky?

WHAT YOUR EARWAX SAYS ABOUT YOUR ABILITY TO DRINK ALCOHOL

If you happen to have the dry type of earwax, you likely had ancestors who lived in east Asia and were not big consumers of alcohol. I've written about why certain human populations have genes that allow them to drink more alcohol than others in much greater detail in my first book, *Survival of the Sickest*.

The short of it is that once people began to live together in larger towns and cities, the water supply became contaminated with pathogenic microbes. Simply put, if you kept drinking that dirty water, you'd die of some nasty water-borne infection. This left your ancestors with two simple survival strategies:

1. Either drink weak or diluted alcoholic beverages that were not contaminated

    or

2. Boil your water and make tea

Which strategy your genetic ancestor employed back then determines how much alcohol you can safely consume today. Those who chose the second survival strategy

---

[1] Please note that regular use of a cotton swab to clean your ears is not recommended. For the purposes of this genetic self-test, please use extra caution when carefully placing the tip of a cotton swab in your ear.

and made tea ended up with genes that were not great at dealing with alcohol and very likely also had dry earwax. Because this group also likely inherited the *ALDH2*2* gene, there's even more reason to limit alcohol intake. In fact, it's been predicted that in genetically at-risk individuals, it's possible to reduce the number of squamous cell esophageal cancers by half just by reducing the amount of alcohol you drink. Yet it's not only the *ALDH2*2* gene that may be predisposing some people with dry earwax to higher rates of cancer, because there are other genetic differences between the dry and wet earwax types.

This is why I would prefer that if you have dry earwax, you cover your genetic bases and not drink any alcohol at all. But I realize this might not be realistic for everyone, so if you're going to choose to drink alcohol, my bottom-line recommendation is that you limit your consumption to no more than three alcoholic drinks a week if you're male, and two if you're female.

And that's absolutely not a license to binge; don't use up your alcohol allowance in one shot! You need to spread out those drinks over the week, as binge drinking is even more risky for those with the genetic background that correlates with dry earwax.

As a final motivator, there has even been some research—presented at the American Society of Human Genetics annual meeting in 2015—that linked high alcohol consumption with signs of increased genetic aging. This means that high alcohol consumption may be prematurely genetically aging all of us.

The DNA Restart Alcohol Intake Guide

Wet-Earwax Type	Dry-Earwax Type
Male: 1–2 drinks per day	Male: 2–3 drinks per week
Female: 1 drink per day	Female: 1–2 drinks per week

But if you've discovered that you have the wet earwax type, your ancestors gave you the genetic know-how to handle larger amounts of alcohol with less-toxic effects. So to maximize alcohol's apparent cardiovascular protective effects, you're free to have two drinks a day if you're male and one drink a day if you're female.

The reason I am advising that women drink less alcohol than men, regardless of earwax type, is because of the increased production of estrogen that occurs as a result of consuming alcohol and its possible connection with promoting breast cancer.

THE DNA RESTART TAKEAWAYS

1st Pillar: Eat for Your Genes

1. Do the DNA Restart Cracker Self-Test.

   Discover your optimal DNA Carbohydrate Consumption Category. The categories are as follows: Full, Moderate, or Restricted, depending upon the results of your genetic self-test.

2. Use your optimal Carbohydrate Consumption Category: Full, Moderate, or Restricted to determine your weekly carb cost allowance.

   Consult the charts on page 17 for your individualized dietary breakdown of fats, proteins, and carbs.

3. Remove all soft drinks and vegetable or fruit juices (not including your daily lemon/lime juice, which we'll discuss in the 2nd Pillar) from your diet.

4. No sweeteners (artificial or natural) allowed, except for a maximum of 2 teaspoons of honey a day.

5. Purge all emulsifiers from your pantry and life.

   Consult the DNA Restart Emulsifier-Busting List on page 35.

6. Eat fermented dairy products.

   If your DNA lets you have fermented dairy products like yogurt and kefir, you should do so at least once a week.

7. Get your iron levels checked if you're male or a postmenopausal woman.

8. Limit red meat and ban processed meats.

   Reduce red meat to no more than two servings (2 to 3 ounces each) every week. No processed meats ever.

9. Do the Cotton Swab Alcohol Intake Test.

   Find out and adhere to your optimal weekly alcohol allowance.

PART II

THE DNA RESTART
2nd Pillar

Reverse Aging

A t 78 years of age, Robert looks much younger. Since no one can guess his real age, he's quick to proudly pull out his wallet and produce his driver's license.

But if you take a close look at the photo on his license, which was taken only 5 years ago, you might think that this ID belongs to someone else. And that's because Robert's transformation only began about 6 months ago. His wake-up call was a minor heart attack from what he described to me as the result of "an entire life of sitting on my butt!"

I actually first met Robert more than 15 years ago. I was conducting neurogenetics research and Robert was an amazingly resourceful volunteer who helped recruit many of the healthy seniors who agreed to participate in some of my genetic research studies. Realizing after his discomforting brush with death that he'd been given another lease on life, Robert seized the opportunity to try to make some real, albeit difficult, lifestyle changes.

The first thing he did was to send me an e-mail with a subject line that read: Payback Time. When I called Robert back, the first thing he said was, "Doc, remember who got you all those volunteers who gave you their DNA for your genetics brain research?"

"Of course I do. How could I possibly forget?" I replied.

"So, can you help me? Now that I've had a heart attack, I want to turn things around. I'm just not sure there's still time."

So I proceeded to put Robert on the DNA Restart. It didn't take long for him to get hooked.

"I love being your guinea pig!" was his initial feedback after he started to apply the latest science and genetics research to his life. The first thing Robert did was the genetic self-tests prescribed in the 1st Pillar, which allowed him to precisely tailor his carbohydrate, protein, and fat intake for his genes. But what he was most enthusiastic about was the exercise plan outlined in the DNA Restart. Robert began a program of at least 10 minutes of high-intensity exercise (appropriate for his age) three times a week and another 10 minutes of strength-based resistance training three times every week. He diligently alternated the high-intensity exercise with the resistance training every day, leaving him with only 1 day "off" (exercise-free) per week.

Since there was a gym attached to a local community center not too far from where he lived, he started to conscientiously either swim or use a treadmill there for the high-intensity component every other day. As for the strength-based resistance exercises, he found a local fitness instructor who volunteered to help for the duration of Robert's DNA Restart. After 28 days, I asked Robert to describe what the experience was like.

"The biggest surprise was my skin . . . [I had gotten] used to it sagging and flapping in the wind, you know. So now I notice a difference when I shave. My skin isn't loose like it was before. It's springier."

As if I didn't believe what he was saying, Robert reached over at this point in our conversation and made me physically tug the skin on his neck.

"It's also the small things I started to notice at first," he recounted to me. "I had stopped trying to open jars or lift anything heavy on my own the last few years. I just didn't have the strength anymore. After going on the DNA Restart, I now have the strength to do things I thought weren't possible. My sleep is also better. . . . You know, I'm going to be 79 next month!"

DNA HEALTH AND LONGEVITY

The latest complex genetics research findings can be distilled quite succinctly—the better you take care of your DNA, the longer you will live.

Over time we all accumulate some types of damage to our genetic material, and once that happens, there's absolutely no going back, or so it seemed. This genetic degeneration was thought of as an inevitable and unavoidable consequence of life.

To the contrary, we are now discovering that it's even possible to reverse genetic aging, and I will be sharing with you the latest research and results from my own work to help you start taking better care of your genetic inheritance. That is the goal of the 2nd Pillar: to get you started on turning back the hands of your own genetic clock.

To accomplish that, we're going to have to do two things. The first is to prevent as much DNA-aging damage as is scientifically possible to date, and the second is to powerfully activate your body's own innate antiaging systems.

When it comes to genetic aging, the balance is between damage and repair. Promoting repair is just as important as preventing the damage in the first place. And the reason why promoting DNA repair and preventing damage is critical is because the more damage to your DNA, *the faster you age.*

Over time we all accumulate some types of damage to our genetic material, and as I mentioned above, this was thought to be an inevitable consequence of life. Thankfully, when it comes to genetics, things are not always as they initially appear. We now know that our genetic code is much more robust, resilient, and malleable than we could have ever imagined.

The bottom line is that if your health goals include weight loss and longevity, you have no choice but to take inestimably better care of your entire DNA package.

I've created an equation to summarize the important concepts that we'll be discussing in this pillar.

[DNA Repair + Antioxidants] - [Inflammation + Oxidative Stress] = Reverse DNA Aging

A - B = C

The goal of this equation (and this pillar) is to significantly reduce DNA aging. "A" represents both the natural antioxidants made in your body and those phytonutrients found in the food you're going to be eating on the DNA Restart. This "DNA repair" is then going to be rocket-boosted by increasing your body's innate ability to repair your own DNA through the exercise program I will be prescribing to you later in this pillar.

All these DNA benefits that you'll be accumulating through "A" would under normal circumstances be naturally reduced by "B"—the amount of inflammation and oxidative stress that your body has been experiencing prior to starting the DNA Restart. But fear not, because I've designed powerful antiaging strategies for the DNA Restart that will help you decrease "B" in a quantum way—which will result in more "C," or reversing DNA aging.

If you're like the DNA Restarters who came before you, you are revving to go by now. But for this equation to work, you have to fervently study and follow the directives in this pillar. Because believe me, when you do, it's going to completely transform how you both look and feel. It takes a geneticist to know that the deepest transformation is always genetic.

The only catch to this whole process is that from this moment on, you're going to need to be on top of all the damage to your DNA. And that's because if you damage your DNA—even in the smallest way—and it's not repaired in the right way or at all, you've just set yourself up for potentially developing a deadly cancer. Or, at the very least, you'll have unwittingly taken another giant leap toward aging.

In the next sections I'm going to show you how to get your body to increase the amount of DNA repair (A), while at the same time lowering the amount of inflammation and oxidative stress (B) you're experiencing.

Stick with this pillar unswervingly and you'll increase A, lower B, and be enviously on your way to reversing genetic aging.

YOUR GENETIC FOUNTAIN OF YOUTH IS DNA REPAIR

Once you start to understand the power you have to control your genetic destiny, by actualizing DNA repair and reducing both chronic and acute inflammation as well as oxidative stress, you'll be even more motivated to make all the changes outlined in all of the five pillars of the DNA Restart.

As I mentioned earlier, like all things in life, there's a balance between damage and repair. I'm restating this here because it is counterintuitive and important. You may be like most of my patients who assumed that reducing all DNA damage is the goal, and that repairing over time, until you reach perfection, would be the zenith of DNA health and longevity. *I'm going to be showing you in this second pillar that both damage and repair are actually essential in elegant balance for optimal genetic health.*

One of the easiest ways to reduce DNA damage is to limit the amount of unnecessary chronic inflammation as well as oxidative stress you might be experiencing in your life. Chronic inflammation affects not only the physical macro structures of the body, such as your skin, joints, gut, and brain, but, of course, all of the DNA contained within your cells as well.

Typically, chronic inflammation is also associated with an elevation of inflammation markers in your blood such as *tumor necrosis factor* (TNF), *C-reactive protein* (CRP), and IL-1B and IL-6. These immune-signaling proteins can have not only an incredible impact on the way your DNA behaves but also on the increased damage it sustains—damage that then leads to aging.

Telomeres = DNA Protectors

Telomeres are physical structures that are like bumpers that protect each of your 23 pairs of chromosomes. As you age, these structures naturally can shorten—the length of your telomeres has been described by scientists as one indicator of your genetic age. And when those inflammatory markers that I just mentioned are elevated because of heightened inflammation, so is the damage to the length of your telomeres. What's incredible, though, is that if you change your levels of inflammation and oxidative stress through key dietary and lifestyle changes like what I'm prescribing through the DNA Restart, you can slow down or even reverse the signs of genetic aging, including telomere length. And that's the key to this 2nd Pillar of the DNA Restart: knowing that you have the power to change your genetic destiny and reverse genetic aging. This knowledge will inspire you to keep focused and disciplined through the next 28 days of your DNA Restart. In the next section I'm going to tell you about the most powerful way to turn on the body's natural system for DNA upkeep and repair (the "A" of the equation).

CHAPTER 10

Exercise Your Way to a More Youthful Appearance

I t's been estimated that every single year, more than 15 million Americans get some type of cosmetic procedure done. And the global market for medical cosmetic services and surgeries has been valued at a staggering annual $20 billion. That's a heck of a lot of money, and it includes many maintenance procedures that need to be repeated regularly, such as injections of Botox and collagen fillers, because they eventually wear off. What's more, this market is predicted to keep on growing, topping out at $27 billion by the year 2019.

There's a really good chance you have already had one or two small medical cosmetic procedures or maybe even gone under the knife for the sake of beauty. Or maybe you haven't yet, but you know someone who has, and you've been considering it for yourself.

However, looking younger can come at a significant price that's not only financial. The biggest drawback with most of the current interventions is that they're medical procedures. And as with any medical procedure, they're not just expensive; they also have their associated risks that can leave you with permanent results that you may find less than appealing. The other issue with some cosmetic procedures is that they often work too well, leaving you with a face that looks nothing like natural beauty should.

But what if I told you that there was something you could do, starting today, that would guarantee you will look and feel younger? Would you believe me if I told you that, beginning today, you can actually naturally increase the amount of collagen in your skin as well as reverse some of the age-related changes that have seemingly irreversibly occurred? And would you believe that it is actually possible to turn back your genetic clock and not just make your skin appear younger, tighter, and more vibrant—but actually lower your risk of becoming susceptible to certain cancers?

Want to sign up for that?

EXERCISE YOUR WAY TO YOUNGER SKIN

Some of the best scientific research being generated by leading experts in their fields today now conclusively proves that there is a totally natural way to make your skin younger and reduce your chances of getting cancer. And it doesn't involve injections, fillers, surgery, or any pills or creams. The best part is it won't even cost you a penny.

This isn't science fiction, but a scientific reality—as research is now showing it's possible for you to reverse genetic aging. The only major hurdle to overcome is yourself. You see, it's not always easy to change your habits, and I should know, because it definitely wasn't for me.

Finding the time and even the energy to keep working at a brand-new lifestyle regimen can feel like a losing battle sometimes. But as the latest research has shown (which I'll be covering more extensively below), you actually don't need hours to get the maximal benefits from certain forms of exercise. Only minutes are required to get your genes to start transforming your body back to a younger version of itself. Got that? The exercise plan I'll be mandating for every DNA Restarter will make your genes work more optimally and your skin appear more youthful.

Frankly, that should be all the motivation you need to get and keep you going. If you are able to commit to making these changes to your life, you will quickly learn that they're not cosmetic but seismic. Don't be surprised if people start commenting about your physical appearance as you transition through your 28-day DNA Restart. That's just one of the perks of reversing your genetic age that you'll have to learn to deal with.

After having my two kids and putting weight on during both pregnancies, I just felt embarrassed and self-conscious going to the gym. Even though I've lost some weight since then, I never went back to feeling energetic like I did in my thirties. Starting to exercise made the biggest difference for me. I started feeling younger and had more energy almost right away. Now I set an example for my two teenage daughters about the importance of taking care of your DNA not just by eating right but exercising as well. Alternating between days of weights and spinning just makes so much sense. It helps to keep me from getting bored. And the best part is that my skin looks so much better! It's amazing to think about all of the changes happening in my DNA. I love it!

—Nicole, 50

GET A DNA RESTART BUDDY ASAP!

An important tip: To ensure optimal success, I highly recommend partnering with someone else to do the DNA Restart so that you're not making these changes on your own. This will make a world of difference for your chances of success. I have seen this time and again. Trust me, it can be the difference between your succeeding or failing.

Research has consistently shown that when you're making big life changes that involve diet and exercise, you're more likely to succeed if you don't do it alone.

I'm prescribing the exercise portion of the DNA Restart for the duration of the 28 days because the research that I'll be outlining later in the chapter is clear: It takes 4 solid weeks for your skin cells to start responding and changing their previous cellular aging patterns. Your genes likewise take 4 weeks of around-the-clock biological reparative work in response to the DNA Restart exercise you'll be consistently doing for the physical changes to really start taking effect. What's incredible is that on a genetic level, the dietary and physical exercise choices that you will be making over the next 28 days (and that I'll prescribe for you in the next section) will be changing the way your genes function. Most importantly, if you're psyched about your new life post–DNA Restart, evidence that I'll be introducing in the later sections of this pillar suggests that these interventions can extend your life by 4 to 6 years.

It's never too late to start! Some of the participants in the research studies that I delved into when designing the DNA Restart exercise indications were in their seventies and eighties. And they still experienced positive changes! Thankfully, we now know that you don't have to be a slave to your DNA, or your biological age for that matter, but rather you can use it as a jumping-off point to start feeling and looking younger.

OKAY, BACK TO THE SCIENCE

When I was a medical student, there was a general consensus that exercise was definitely good for your heart. It's incredible to see how far we've come in our scientific understanding of the potential that specific exercises can have to completely transform us physically. Most excitingly for me, these transformational changes are now being shown to also apply genetically. Exercise has been implicated in decreasing your risk of developing such health issues as neurological disorders, diabetes, cardiovascular disease, depression, and even cancer. There's even research that shows that exercise can lengthen your life.

So let's dive in and take a quick look at the fascinating science behind what

Robert, whom you met at the beginning of this pillar, was experiencing. And most importantly, I'll explain what you'll need to start doing from the first day of your DNA Restart, so that you can start seeing and feeling all of the positive effects of reversing your genetic age.

To do that, I need to introduce you to something that's been keeping you alive from the very beginning of your life. Meet your mitochondria.

MITOCHONDRIAL DNA AND AGING

These specialized, energy-producing structures are not only the key to your survival but also the key to keeping you looking and feeling young. Mitochondria are found within each and every one of your cells, except for your mature red blood cells.

What's really special about mitochondria is that they have their own strand of DNA—and it's quite different from your three-billion-letter genetic code. This "mitochondrial code" is the smaller sister code to your genomic code. It's much more compact and rather sensitive to oxidative stress—so much so that we all inherited special genes whose job it is to make sure that our mitochondrial DNA is kept safe.

Mitochondria are thought to have been originally free, independently living microbes that at some point in the evolutionary history of animals on this planet joined forces with more complex cells. So today, in exchange for helping you produce energy from food more efficiently, mitochondria get a safe place to live inside almost

DNA Restart Health Tip #8

Find an all-out intermittent exercise that you like doing and perform it three times a week. Here are a few examples of activities you can try.

1. CrossFit
2. High-intensity anaerobic interval training (aka HIIT)
3. High-intensity cross-training (aka HICT)
4. High-intensity dance class
5. Running
6. Spinning

all of your cells. Most importantly, they get constant maintenance and upkeep of their own mitochondrial DNA as well, compliments of you as an organism.

For example, you possess a specific gene within your DNA, called *POLG1*, whose job it is to work just like a copy editor or proofreader, making sure that mitochondrial DNA doesn't end up with genetic "spelling" mistakes. And this is what helps keep mitochondria functioning optimally, which is also darn good for you.

The other thing that your cells do is work hard to protect mitochondrial DNA from oxidative stress, and they do this by having enzymes that help mop up any chemical trouble in your cells and body before it gets out of control.

In fact, one of the major theories of aging posits that mitochondria are the crucial linchpin in keeping you healthy and, most importantly, alive. The decline in your mitochondrial DNA and their subsequent health is what eventually leads you to age. Not having enough mitochondria, as well as not having them work optimally, is like having your local power plant start failing and shutting down. Everyday conveniences like kitchen appliances, home lighting, heating, and air-conditioning all rely on a constant supply of electrical power. As some of you may have experienced during the electricity blackouts during Hurricane Sandy, it's very hard to lead a modern life without electrical power.

And this is what is thought to happen when your mitochondria start failing or not working efficiently; every cell in your body starts running out of power and begins to fail. Essentially, you start to age.

There are even mice that have been bred to have mitochondria that stop working early because they have been given a *POLG1* gene that doesn't work well. Without a proper version of this gene, the mice get loads of spelling mistakes in their mitochondrial DNA, which is something that happens to humans as we start aging. Because of all these spelling mistakes, they also get all the usual signs associated with aging, but long before their time.

The mice experience premature ailments including cataracts, hearing loss, loss of muscle mass, hair loss, and constant joint pain—all because of some simple spelling mistakes that occurred in their mitochondrial DNA. And this sets off a devastating sequence culminating in every organ in their furry bodies starting to fail.

WHAT MICE TEACH US ABOUT EXERCISE AND AGING

To learn more about this scientific work on health and aging, I visited one of the world's leading experts on the fascinating role of mitochondria in health and disease, Mark Tarnopolsky, MD, PhD, at his research laboratory at McMaster University in Hamilton, Ontario, Canada.

Actually, as Dr. Tarnopolsky recounted to me, he never planned on pursuing a career as a physician-scientist. As a child, he was, as he describes, "rambunctious" and relied heavily on sports to help him focus and get all that energy out productively. Because his first love was exercise, he naturally thought that a career as a high school physical education teacher would let him constantly stay active and in great shape.

But while he was attending college, his love of exercise was supplanted with a burning and insatiable desire to understand why exercise always left him feeling so great. This shot him off in a completely new vocational direction that saw him complete both an MD and PhD, and then specialize in trying to systematically understand the robust effects that exercise could have on the body.

So when he asked a colleague if he would provide him with a few of his mice that had lost the ability to correct their mitochondrial DNA spelling mistakes because of a broken *POLG1* gene, most people would have wondered what he was proposing to do with these mice. If they were to find out that he asked for these specific *POLG1* mice because he wanted to exercise them, they would have thought he was just a little crazy.

That's the same gene I was telling you about earlier that constantly works hard to fix problems in the DNA within mitochondria. Without a working *POLG1* gene, these poor mice become very old long before their time. This is thought to happen because their mitochondrial DNA becomes full of errors and mistakes, which, as I explained previously, triggers all of the symptoms of aging. What Dr. Tarnopolsky wanted to see, which was pretty much heretical at the time, was whether or not forcing these mice to exercise three times a week for 30 minutes would improve their condition.

Actually, if you were to ask most medical researchers, only just a few years ago, they would have predicted that these mice with poorly functioning mitochondria would actually be worse off after all that exhausting running on a treadmill. Imagine forcing a fracture-prone, frail, and elderly grandparent to walk briskly on a treadmill three times a week. You'd think they'd probably end up with mangled joints or suffer a heart attack, or both or even worse. You might even guess that they'd probably die sooner. And you'd be wrong.

Not many people paid much attention or had high hopes for Dr. Tarnopolsky's exercise study at the time. Thankfully for all of us, Dr. Tarnopolsky didn't pay attention to scientific dissenters. Instead, he forged ahead and exercised his *POLG1* mice.

Taking the *POLG1* mice he was given, his team divided them into two groups: one that he exercised three times a week on a treadmill for 4 months, and a second group that got to lounge around waiting to get old long before their time.

And then the miraculous happened—which is saying a lot in the world of research science.

The mice that ran did not die sooner or fare any worse because of the exercise they were subjected to, as most people would have predicted. In fact, the opposite happened. It was as if the exercise rescued, not damaged, their mitochondria that were prone to getting errors in their DNA. For all intents and purposes, the exercised mice looked totally youthful and vigorous as compared to the group of mice that just sat around and became partially bald and hearing impaired.

What's behind this amazing transformation?

Dr. Tarnopolsky and his team discovered that strained muscles that have been exercised release certain molecular signals, such as interleukin-15 (IL-15) and others, which he named *exerkines*. These signals cause systemic benefits that essentially stop and reverse the aging process. Most importantly, exerkines stopped the genetic aging process in these mice. They looked just as young and healthy as genetically normal mice. And interestingly, they also were less socially isolated and more fertile. No doubt, this was an unexpected bonus side effect to an already growing body of positive evidence Dr. Tarnopolsky was generating.

According to Dr. Tarnopolsky, exerkines are thought to drive the same process in all of your human cells as well, in two important ways.

First, exerkines drive your cells to make more mitochondria, to replace the frail ones that are burgeoning with mutations in their DNA. This is like getting a power plant upgrade in every cell in your body that has mitochondria, including the brain.

Second, exerkines are also driving your own cells to proactively prune and get rid of mitochondria that have stopped performing well—like a gardener who prunes a diseased branch off a sick tree, which speeds up the tree's overall return to health.

Since these game-changing discoveries, Dr. Tarnopolsky hasn't spent much time resting. He's been very active as a physician-scientist doing research, and he definitely hasn't stopped exercising either. His most recent study, published in 2015, compared people who actively exercise (completing a high-intensity workout four times a week or more) to people who are sedentary. What his research found was that not only did the exercise group have more mitochondria in every cell—that's the power upgrade I was telling you about—but their genetic health was also more intact, and their skin appeared structurally younger. This further substantiated his previous research that exercise over the years can help keep you genetically younger by resulting in more mitochondria that are free of DNA errors.

So to find and develop the best exercise strategies for your DNA Restart, I asked Dr. Tarnopolsky what were the most important tactics that you could start employing today to reverse genetic aging and improve your mitochondrial health.

He cited the exact study I just told you about as a small but very significant example of how starting and maintaining an exercise program can have such a powerful effect on your overall DNA health.

He went on to explain how instead of just ending their research after finding out that active mice have more mitochondria and younger-looking skin, he and his colleagues decided to take things one step further. They forced a sedentary group of older adults to do endurance exercises for the next 3 months.

What they discovered at the end of the study should be of special interest to anyone who wants to look and feel younger. And that's because after only 3 months of exercise, the formerly sedentary group had skin with much more collagen in it than it had before, as well as measurably more mitochondria in all their cells. It's as if through endurance exercise alone, they reversed some of the signs of normal human aging, like sagging and less-taut skin, for example, that usually happen to everyone as they age. And it's the same exerkine, IL-15, that was responsible for these desirable changes. Yet what's important was not the overall amount of IL-15 that was being released after exercising, but that it was *pulsatile.*

The level of IL-15 went up after exercise and then it came down. It's these natural *pulses* of IL-15 released from stressed and strained muscle that then command the rest of the body to reverse genetic aging. And when these pulses of IL-15 are spread out over an individual's day and week, they initiate the dramatic age-reversal changes.

Dr. Tarnopolsky hypothesizes—and I agree with him—that having fewer of these IL-15 pulses during the day is what is driving aging in the body, especially the part of aging that's most visible to everyone else: your skin.

It's too early to know for sure if and when Dr. Tarnopolsky will be awarded the Nobel Prize in Physiology or Medicine for some of his truly groundbreaking work showing the power of exercise to reverse aging, but why should you wait?

That's why I'm going to prescribe for you the most important strategies you can employ during your 28-day DNA Restart and beyond to take advantage of your body's powerful ability to heal, weed out damaged mitochondria, and keep you both looking and feeling younger.

YOUR DNA RESTART EXERCISE PRESCRIPTION

Here's the simple and powerful two-pronged approach to DNA Restart exercising. First, you need to choose any exercise you find enjoyable that gets you moving and your heart rate up. The goal is to mimic the type of "eat or be eaten" scenario your genetic ancestors likely experienced sometime in your evolutionary past. Think about it for a minute. If your genetic ancestors could not chase down and kill their

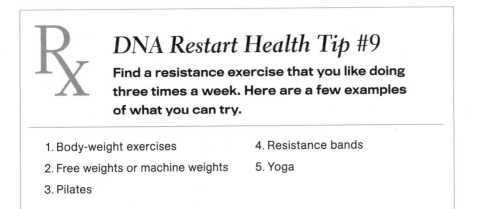

DNA Restart Health Tip #9

Find a resistance exercise that you like doing three times a week. Here are a few examples of what you can try.

1. Body-weight exercises
2. Free weights or machine weights
3. Pilates
4. Resistance bands
5. Yoga

---

prey, or run away from another animal that was trying to turn them into dinner, how long would they have lasted?

Probably not long at all. That's why even today your body still uses IL-15, released when you are intensely working out, as the signal to stay younger for longer and help your clansmen out. If you stop moving intensely, what you're telling your body is that it's time to stop being a burden on others—it's time to die.

I want you to live a long and more youthful life. So let's get back to what you're going to need to do.

1. Choose any exercise that you find pleasurable. (If you think there isn't one, you just haven't looked hard enough. So look again and find one.) Do that exercise intensely for as little as 3 minutes, with a 10-minute warmup and 10-minute cooldown period, three times a week. In DNA Restart Health Tip #8, I've provided you with a prescription pad full of activities that can qualify.

   To calculate your maximum heart rate, take your age in years and multiply that number by 0.7 and then subtract that value from 208. For instance, if you're 43 years of age, then multiply that number by 0.7, which is about 30, and subtract that from 208 to get a max heart rate of 178.

   If you like equations, here it is:

208 − (0.7 x age in years) = Maximum Heart Rate

   To make sure that you're working hard enough to reach your health goals when exercising, you should be aiming for a minimum heart rate that is at least 70 percent of your maximum heart rate that you just calculated. To get that number, multiply your maximum heart rate by 0.7.

2. Find a resistance, strength-based exercise that you can do three times a week, even if it's only at home. We're talking about another 15-minute time commitment to get your genes activated and keep your body looking and feeling not only younger but stronger. If you don't feel comfortable starting on your own, you can do it together with a fellow DNA Restarter, or use a trainer to get your own set of exercises and make sure that you're doing them right. In DNA Restart Health Tip #9, I've also provided you with a prescription pad full of activities that can qualify to give you some ideas. Remember Robert? In his late seventies, he had only 1 day off (exercise-free), and the same goes for you. You must be alternating each day between strength-based resistance exercises and all-out intermittent exercise. No whining—you and your mitochondria will thank me.

CHAPTER 11

Taking Your Genetic Age Reversal One Step Further

What I've created in the DNA Restart 2nd Pillar, Reverse Aging, is the perfect plan for you to mitigate and reverse the genetic aging process and get your body to do what it knows how to do best, and that's repair your DNA. There are lots of clinically proven ways to simply "lose weight" in the short term, but what's the point if you're not going to take care of your DNA? *Failing to care for your DNA means that you're literally shaving years off your life span, and by doing that, you won't have the time you rightfully deserve to live.*

The next component of this pillar will involve getting you to stop needlessly breaking your DNA. Broken DNA, in case you're curious, promotes genetic aging and eventually cancer.

As you recall from the 1st Pillar, Eat for Your Genes, many of the genes you've inherited play a big role in whether you can healthily metabolize your food or not. To find the best types of foods and spices that address your genetic uniqueness, I've spent the last 2 years traveling to very distant parts of the world to research and discover what you should be consuming every day for maximal genetic wellness throughout your 28-day DNA Restart.

You'll be glad to know that one of the things I discovered was that some of the most powerful phytonutrients on earth—those that best help preserve and protect your DNA—are not located in distant lands, with labels that you cannot even begin to pronounce. They're actually sitting in your kitchen already. You just need to know how and when to use them.

So let's dive in!

STOP BREAKING YOUR DNA

Before I begin prescribing powerful ways to mop up oxidative stress and reduce inflammation, let's spend a little time discussing the DNA aging equation I introduced you to earlier. Just to jog your no doubt excellent memory, here it is again:

[DNA Repair + Antioxidants] – [Inflammation + Oxidative Stress] = Reverse DNA Aging

A – B = C

Alas, lots of things in life break our DNA. Some things you may be aware of, and others may surprise you. Staying out too long in the sun or taking a transatlantic flight both expose you to radiation that can damage your DNA. The airline industry might not want you to know this, but every time you take a long-haul flight, your body has to work hard to not only compensate for dehydration but also repair the genetic damage you experienced while you were soaring high above the earth. All that frequent radiation exposure has poked many holes in your genome and has sped up the aging process. The higher up in the atmosphere, the longer you travel, and the farther you are from the equator, the less protected you are from cosmic and ultraviolet (UV) radiation and the more damage your DNA will sustain. This is likely why cabin crews have been found to be at twice the risk for melanoma, the deadliest type of skin cancer.

Exposure to UV radiation is not new for humans. It helps explain one of the most obvious genetic differences among us—our skin color. The darker your skin, the more your ancestors evolved to be protected from the DNA-damaging UV rays of the sun. This is also why, conversely, the lighter your skin color, the more you need to avoid exposure to UV radiation during its peak intensity time (usually 12:00 to 2:00 p.m.) during the day, as well as cover up as much as possible and use a broad-spectrum (UVA + UVB) sunscreen.

We've also all inherited very different genes that help to repair DNA damage. Some of us are better skilled genetically at cleaning up a biological mess and mending the damage done by daily living. I've spent many years doing research and working with patients to improve the human body's ability to cope with the many dangers our DNA faces every single day, with the ultimate goal of extending life and improving health.

This work has culminated in a large global multicenter research trial involving patients who are born with a reduced ability to repair their DNA. Since these patients almost always develop cancer from their inability to repair their DNA efficiently, I proposed that perhaps taking a specially formulated antioxidant daily would improve their genetic health status. I was fortunate to collaborate with other experts in my field to design an international research trial that is currently investigating whether my proposed intervention could actually improve the genetic health of these patients. While results of this study are still a few years away, there are things I've uncovered through my research that you can start doing right now to begin to take better care of your DNA and, most importantly, reverse genetic aging.

The promising news is that our DNA repair system is really very hardy. This means that we have the potential to repair and reverse many of the genetic aging events that until a few years ago we thought were unavoidable. Just give our genome the right tools and time and let it do what it knows best.

Of course this doesn't mean that we have completely outwitted aging, but rather that we now know with scientific certainty that we can all play a tremendously active and important role in the outcome of our lives.

Even monozygotic, or "identical," twins who start their lives with practically the same DNA do not always age genetically in the same way. Every day there are new scientific studies in the field of epigenetics that are showcasing how everything we do in our life right now matters. What I've distilled from my own research and work with patients over the years is that the foods you eat and how much exercise you demand of yourself every week, taken together, are the single most crucial factors when it comes to transforming your DNA and the way it behaves. *Food and exercise are the most powerful natural tools in your medicine cabinet.*

This ultimately means that you have the power in your hands right now to completely change your life, genetically speaking.

Start to properly feed and exercise your genes, and your body will thank you for it immensely. And you'll start to notice the changes physically, mentally, and even emotionally over the next 28 days and beyond as you and your genes transition through and complete your DNA Restart.

So let's get started reversing the damage done by years of needless and mindless lifestyle habits that have damaged and aged your DNA. But first, there's something very important that needs clarifying. It's so important, in fact, that I'm going to give it its very own subtitle:

A HEALTHY WEIGHT ≠ HEALTHY DNA

As I mentioned briefly earlier, getting thin is something lots of people can do. But news flash: Skinny does *not* automatically mean healthy, certainly not from a genetics perspective. Losing a healthy amount of weight, keeping it off, and most importantly, getting your genes thriving is the primary goal of the DNA Restart.

Many proponents of high-protein or exclusion diets have no idea how to take your own personal genetics into account. And, unfortunately, that means many people have been told to limit the exact foods that your DNA requires to keep you genetically young and cancer-free.

That's why I do not advocate limiting and excluding the very foods that have been consumed by your various genetic progenitors for millennia. The foods that I'm recommending on the DNA Restart are instrumental in protecting your DNA and keeping you and your mitochondria young. The only way to really protect your DNA is never as simple as just eating more blueberries.

Just like you learned in the 1st Pillar of the DNA Restart, Eat for Your Genes, regarding the amount of iron intake from red meat, what spells out health for one person can severely harm another.

The best part of delving deeper into this 2nd Pillar is that you're not only going to start feeling better outwardly, but inwardly as well, as your DNA will start to function optimally.

WHAT OXIDATIVE STRESS MEANS TO YOUR DNA

By now you've likely heard all about oxidative stress and the damage it can cause the body.

That damage is what you definitely want to avoid happening to your body and especially your DNA, because oxidative stress and inflammation have now been implicated in many chronic diseases such as diabetes, cardiovascular disease, osteoporosis, Alzheimer's disease, Parkinson's disease, and, of course, cancer.

When people conjecture about the causes behind oxidative stress in our bodies, they usually think about air pollution, chemicals in our environment, and ultraviolet radiation. As we discussed in the 1st Pillar, Eat for Your Genes, oxidative stress can also be a completely normal by-product of your own body's day-to-day functional processes, such as turning food into energy.

For example, along with being an excellent source of protein, red meat has many naturally present toxic components, especially iron, that can literally rust your body, increasing oxidative stress and damaging your DNA, every single time you sit down for a steak dinner. And then there are the many other toxic compounds that are normally produced every time your body digests and converts any food into usable energy. Oxidative stress is often the not-so-simple biological cost of living.

Few people want to live near a coal-fired power plant for well-founded fear of toxic exposures, but the cells in your body don't have that choice and have to deal with the chemical oxidative onslaught that you've just created by your sometimes faulty dietary choices.

MOST OXIDATIVE STRESS IN YOUR BODY IS "NATURAL"

You may be surprised to discover that not all inflammation is bad for you. Sometimes it's completely beneficial, because it's a normal pathway your body uses to repair itself, just like you learned about what happens every time you exercise strenuously. There are many times in your life when oxidative stress has literally saved your life. Oxidative stress is actually used by neutrophil cells, as a part of your body's immune system, to kill invading microbes and rogue cancer cells.

The problem for your body, and especially your DNA, is when it gets too much of a "good" thing. An example of this is the superoxide anion (O_2^-), one of the human body's more common free radicals that are produced naturally. There are also other naturally unstable chemical compounds such as nitric oxide (NO), nitrogen dioxide (NO_2), and hydrogen peroxide (H_2O_2) that are produced and found within the body.

The cell membrane, the living bag that surrounds each of your cells, is particularly rich in polyunsaturated fatty acids that are easily damaged by these very reactive chemical compounds, known as *reactive oxygen species*, or ROS. It's also thought that these ROS, and the processes they drive, can react with fats that are circulating in the blood and essentially make them "rancid." Once that happens, the rancid fats can trigger inflammatory reactions that are thought to fuel disease processes such as atherosclerosis, or hardening of the arteries.

Most importantly, though, it's the damage that can occur to DNA from too much ROS that is particularly worrisome, as it's kind of like the express lane on the highway to mutations, premature aging, and cancer.

Thankfully, the body latently has the ability to deal with a normal amount of oxidative stress. Your body accomplishes this by making its own antioxidants, which work like fire extinguishers to put out the oxidative stress fires that may be raging in your body. Whatever antioxidants your body doesn't make, or doesn't make enough of, you better make sure that you're getting from your diet.

The problem begins when we substantially increase the demand on our bodies to detoxify greater amounts of oxidative-stress-producing chemical pollutants from the food we eat, compounded by the other chemical pollutants we're exposed to in daily life. When this is coupled with unnecessary inflammation caused by ingesting emulsifiers, for example (as we discussed in the 1st Pillar), our bodies struggle to cope with the onslaught of oxidative stress. It's like not having enough fire extinguishers on hand and then a small fire becomes bigger, to the point that it's no longer possible to contain. When this happens, free radicals are unleashed to roam unhindered, harming our bodies and ultimately damaging our DNA and speeding up aging.

So too much oxidative stress can lead to DNA no longer functioning properly. And not enough puts you at risk of infections and cancers. The bigger problem in the long run is damage to your DNA. Do you know what happens when your DNA stops working? Every biological process in your body stops, too. And you don't want that—after all, you and your DNA have history.

AN ODE TO YOUR DNA AND HOW MUCH IT'S DONE FOR YOU

Your DNA actually predates you. It was created years before you were born, while your mother was still in *her* mother's womb. That very strand of your mother's DNA braided with a strand provided by your father, and you were eventually created into the baby version of you who was born any number of years ago.

But for our purposes today, trust me when I say that without a perfectly coordinated and working genome, eventually what your body did with ease (digest your dinner, play piano, remember your grandmother's birthday, etc.) will all go out the window. This includes your ability to make and access new memories. Yes, even the neurological hardware in your brain responsible for producing and storing memories is reliant on your DNA.

Many of the things we do and experience in our lives, such as subjecting ourselves to extraneous oxidative stress because of poor dietary and lifestyle choices, have the potential to harm and damage our bodies and genes. Some of this damage can be directed to the structure of your DNA itself, which would look almost like pulling apart a pearl necklace. When this happens, your body then struggles to pick up all the pieces to repair. As you may have experienced in any home renovation project, sometimes things can go terribly wrong. The same is true when cells in your body work hard to keep your DNA pristine from moment to moment. Your cells are constantly editing and reediting their DNA to keep you healthy and alive.

As we all genetically age over time, minor mistakes start creeping into our DNA until our bodies can no longer keep up with the cost of repairs and rogue cells turn into a full-blown cancer. Even the constant act of putting together your DNA strand can invariably lead to mistakes. This then puts you at greater risk for a cell to malfunction, or tumble out of control, as it would in the development of cancer.

WHY YOUR EPIGENOME MATTERS

Have you met your epigenome? It's also a pretty big deal, biologically speaking. While your DNA was busy expertly coordinating all of the biological events of your life, these events then left their very own annotations, called the *epigenome*, on your DNA as well. Your epigenome is like genetic notations on the margins of a book that your body uses to store information. Many of the things you do in your everyday life can directly impact your epigenome. It's also what distinguishes "identical," or monozygotic, twins from each other, especially as they age, as their epigenome changes to reflect their different life choices.

Some of the epigenetic marks are activated through a process called methylation, where a gene is either turned "on" or "off." Not all of your DNA is used by every cell; it is a three-billion-letter code after all. Your genome has a mechanism where it leaves relevant sections more accessible, just like when you dog-ear a favorite page in a book to remind yourself to go back there. And everything you've ever experienced in your life has left some kind of mark on the collection of genetic material that's been handed down to you from your ancestors. For a much more in-depth and even more fulsome description of epigenetics, you can read my third book, *Inheritance: How Our Genes Change Our Lives, and Our Lives Change Our Genes.*

But suffice it to say here that your DNA does not easily forget.

Your genome "remembers" by either methylating or acetylating proteins called *histones.* Everything you eat, think, and do changes all of these settings on your epigenome. Change the methylation or acetylation status of your DNA that prevents you from developing malignant cells—and cancer you will get. That's why small epigenetic differences are the difference between cancer and health.

The bottom line is that if your health aims are the combination of both weight loss and longevity, then you have no choice but to take much better care of your entire DNA package. And you're about to see how that includes making consistent, genetically discerning food choices the DNA Restart way.

CHAPTER 12

Why DNA Loves Phytonutrients and You Should, Too

You might be wondering why the botanical world is a veritable phytochemical pharmacy. Some of the reasons behind plants' mastery of the chemical arts have to do with the fact that they don't have legs, wings, or tails. Unlike most other organisms on Earth, plants can't get up and run away at the first sign of trouble.

Bad weather, too much sun, poor soil, and the constant dangers of being eaten—all take their stressful toll on plants. While many people get episodic cravings for some vegetable greens, we are by far not the only threat to a plant's life and well-being, as virtually everything and everyone on this planet seem to like having a salad from time to time. Because of this, plants have learned how to defend themselves when they get attacked by animals, insects, fungi and bacteria, and even other plants. To do this, plants have become superbly skilled in the art of poison.

Plants respond to the stressors in their environment by producing an impressive array of phytochemicals, which are like a very large family of natural botanical survival compounds. These physical stressors can include things like being exposed to high amounts of ultraviolet radiation, which can occur when plants grow at very high altitudes. Or a simple lack of water, which can happen on a particularly hot and dry day, can stress a plant.

So over millions of years plants have become incredibly resourceful and supremely chemically smart.

I'm sure you've heard the old dictum "you are what you eat," and it's totally true when it comes to filling your body up with all of the phytonutrients produced by plants. When we eat or drink phytonutrients that plants have made, we're filling our body with their unique genetic and chemical wisdom.

Though ascorbic acid, a form of vitamin C, gets a lot of attention, it is far from the only phytonutrient you get from consuming plants that can have a big physiological impact, protecting both your body and your DNA from microbial, bacterial, and other such invaders. Fruits, vegetables, and spices all contain a rich cornucopia of

R℞ DNA Restart Health Tip #10

Fill your shopping cart with these phytonutrient-packed vegetables and herbs. Remember to diversify and not get too much of any one food. And let your eyes lead you, because the more colors (antioxidant, DNA-protecting plant pigments) you eat, the better!

1. Artichokes	17. Kale (cooked only)
2. Asparagus	18. Lemongrass
3. Beetroot	19. Mint
4. Broccoli	20. Mustard greens
5. Broccoli rabe	21. Okra
6. Brussels sprouts	22. Onions (all varieties)
7. Cabbage	23. Oregano
8. Carrots	24. Parsley
9. Cauliflower	25. Peas
10. Collard greens (cooked only)	26. Peppers (all colors)
11. Coriander	27. Purple or yellow potatoes
12. Eggplant	28. Rosemary
13. Fennel	29. Sage
14. Garlic	30. Scallions
15. Green beans	31. Sweet potatoes
16. Hot peppers	32. Zucchini (dark green variety)

phytonutrients. Of these, there are six key carotenoids that are found within the foods we eat; these are pigment-like antioxidants that are significant to human health: α-carotene, ß-carotene, ß-cryptoxanthin, lutein, lycopene, and zeaxanthin. In the DNA Restart, you're going to get these through a specific combination of fresh fruits, vegetables, nuts, and olive oil that I will be prescribing you to eat throughout the 28 days.

The botanical world's version of a phytochemical self-help guide is contained within the DNA Restart's Fresh Greens Salad with Herbs and Spiced Nuts on page 235.

R℞ DNA Restart Health Tip #11

The selection of phytonutrient-packed fruits listed below will provide DNA Restarters with a wide spectrum of beneficial compounds. The same rule applies here as for vegetables: Try not to have too much of any one fruit, because diversity is essential for your genetic health.

1. Apples (all varieties)
2. Apricots
3. Bananas
4. Blackberries
5. Blueberries (wild when possible)
6. Cherries
7. Citrus fruits
8. Cranberries
9. Figs
10. Guava
11. Lychee
12. Mango
13. Papaya
14. Peaches
15. Pears (all varieties)
16. Pineapple
17. Plums
18. Pomegranate
19. Purple grapes
20. Raspberries
21. Red tomatoes
22. Watermelon

Why are plants so smart? Because they have no choice—it's called survival. And since plants cannot always choose where they happen to set down roots and live, they have to get chemically creative. If that location happens to be at a high altitude, for example, they're going to be exposed to a significant amount of oxidative stress from the increased UV radiation that they're receiving every day. *The phytonutrients we consume from these plants provide us with the knowledge of how to survive on this stressful place called Earth.*

When I traveled to Peru for research and visited the Altiplano plateau, I experienced firsthand what it was like to be one of these plants when I was exposed to too much radiation. With every 3,500 feet I was ascending, I was getting zapped with an additional 10 percent more radiation.

When I reached my destination at an eventual altitude of more than 13,000 feet, I was being exposed to 30 percent more UV radiation than when I started my journey back in the capital of Lima, which is at sea level. My skin definitely paid the price; even with a generous application of an effective broad-spectrum sunscreen, I still got a bad sunburn after being out in the direct sun for only 30 minutes. That's never happened to me before or since that trip to the Altiplano.

PHYTONUTRIENT-RICH SPUDS

The reason I traveled to Peru was to research how plants, particularly potatoes, cope with the stressors of life. Although the potato has been much villainized over the last few years, it can actually be a fantastic source of phytonutrients, as you'll soon see.

There are a bewildering 4,000 varieties of potatoes, and they have a kaleidoscope of phytonutrients, which have been shown to reduce inflammation and quash oxidative stress. Remember what I was telling you about why plants make many of these phytochemicals in the first place—not as phytonutrients for our "healthy" pleasure and consumption, but rather so that they can deal with oxidative stress. Some of this stress is the result of increased exposure to UV radiation—just like what I experienced in the Altiplano in Peru. And UV radiation is powerful and directly damages your DNA by shattering it. This is why too much UV-radiation exposure will eventually cause cancer. Your DNA ages and starts shattering every time you simply sit in the sun!

My guide in the Altiplano was Alejandro Argumedo, who works for ANDES, an internationally recognized nonprofit association that works with indigenous groups to establish Biocultural Heritage Areas like the Potato Park that I visited.

This area encompasses almost 40,000 acres and was devised as a way to maintain the extremely important natural biodiversity of Andes root crops being grown by the indigenous community, including mashua, oxalis, and potato. After all, Peru is where the potato was born, and the Inca civilization thrived on its flesh.

The Potato Park is a veritable ark of botanical wisdom that is held in trust and protected by the Guardians of the Potato—locals who have banded together to protect and pool their biocultural heritage. They have been incredibly generous with their wisdom and openly share seed potato with other communities and researchers around the world.

The diet of the Guardians of the Potato is predominantly—you guessed it— potato, and rarely some protein sources such as guinea pig. They often trade their potatoes with people living at lower altitudes who can supply them with fruits and vegetables that grow easily at lower altitudes but not at the high elevation where the guardians live.

Yet as we were making our way up steep roads and switchbacks in a convoy of three jeeps, it was something that Alejandro told me about the local people's unique health status that really piqued my attention. When I asked about the health of the locals, he told me that the Guardians of the Potato, who live year-round at this high altitude, actually have much lower rates of eye diseases such as cataracts and macular degeneration and even skin cancers. Yet he wasn't sure why.

I initially found this information rather surprising. It's been known now for quite some time that, because of the increased exposure to UV radiation, living at high altitudes can predispose you to more eye damage and skin cancers. I thought about all of this while squinting in the very bright high-altitude sun. I couldn't imagine how the locals might be having fewer cases of skin cancer and eye damage given the constant, year-round extra 30 percent UV radiation from the increased altitude.

All that extra UV radiation is known to increase the amount of damage to DNA by producing structural rearrangements. Just being in the sun for a second or two can usually cause up to 100 dimers[1] to form. And this is in just one cell in your body. Many of your enzymes then need to work overtime repairing the damages to your DNA, which are like kinks in its Slinky-like helical structure. As you may remember if you actually had a Slinky, once they have a kink, they don't work as well. Our DNA is just like a Slinky in this way, and a kink in its three-dimensional structure is the first step in DNA aging, as it can lead to DNA-copying mistakes that can quickly cause a normal cell to become malignant, and then, before you know it, a new cancer is born.

So all this talk of a lot less skin cancer and eye disease in a population that spent almost all of its time outdoors with an extra 30-percent dose of UV-radiation exposure left me rather puzzled. When I asked Alejandro more about this apparent UV-exposure conundrum, he told me about some of the community members who left years ago to find work in the capital, Lima, and never returned. As it turned out, the eyes and skin of these people were not protected in the same way. Compared to the local Guardians who didn't leave the Altiplano, the Guardians residing in Lima had higher rates of skin cancer and eye disease. I asked Alejandro to translate on my behalf and ask one of the Guardians why he thought their skin and eyes are in such good shape.

I was stunned by the poignancy of his reply. Alejandro's translation of his response to my question was that they were not in fact Guardians of the Potato; rather, they believed that it was the potatoes that were guarding them.

This got me thinking, Could it be something in these high-elevation potatoes that the Guardians were predominantly consuming year-round that was protecting their DNA from aging, as well as from skin cancer and diseases of the eye?

---

[1] If your DNA strand is like a necklace, a dimer is like an unsightly "knot" that damages the way your DNA stores information. To be able to function properly, your DNA needs to be constantly repaired of these dimers.

Many varieties of potatoes do, in fact, contain a significant amount of antioxidant phytochemicals, such as carotenoid isoprenoid molecules. The variety of potato that sports a deep yellow flesh, for example, is rich in zeaxanthin, antheraxanthin, and lutein, which are pigment-like antioxidant phytochemicals that are important in the prevention of macular degeneration of the eye.

Many of these phytochemicals are actually produced by plants, such as the high-altitude potatoes, to help them prevent and deal with the increased oxidative stress and DNA damage caused by the extra levels of UV radiation to which they're routinely exposed.

It's still too early to know what gives the Guardians of the Potato more protection from the UV radiation they're exposed to by living year-round at such high elevation. In the meantime, I would still encourage you to get as much of these phytonutrients from your diet as possible.

Interestingly, though, it seems that when plants are grown under controlled conditions, such as what happens in today's modern greenhouses, and they are consequently not stressed by things such as increased UV radiation, these pampered plants contain fewer phytonutrients, even if they are organic.

On the other hand, some organically grown produce that's field grown is, in fact, naturally more stressed because it is less protected from natural predators, so it contains more phytonutrients as a result. Eating this produce, especially when it's prepared the DNA Restart way, will make many more phytonutrients available to you. Ultimately, what you want to do is fill your diet with phytonutrient-saturated foods (I'll be giving you many more examples of these later) that will have a big impact on your health. Many of the phytonutrients that help protect your DNA and reverse its aging are found in the greatest quantities in deeply colored plant foods. This is why I suggest you let your eyes lead the way, and when choosing potatoes, for example, pick the ones with the deepest-colored flesh.

Practically speaking, this also means that on the DNA Restart, you will be choosing field vegetables and fruits, and not greenhouse-grown produce, even if it's organically grown. Stressed produce, such as an organically grown field tomato, will not only have more phytonutrients, but will also taste much more delicious. Many of these phytochemicals are the result of millions of years of botanical evolution and constant upgrading, so why not use them to stop oxidative stress and reverse genetic aging, just like the plants themselves do?

Remember: Phytonutrients = Phytoknowledge, the most natural way to prevent and protect your body from chronic diseases and DNA aging.

CHAPTER 13

The Phytochemical Flip Side: Foods That Harm and Age Your DNA

When asked, most people will tell you that they would rather not have their foods sprayed with pesticides, fungicides, herbicides, and insecticides. And this purchasing choice has spawned a multibillion-dollar-per-year industry called organic foods. What you may not know is that some of the most toxic, carcinogenic, and mutagenic compounds known on planet Earth are all, shockingly, completely natural and therefore organic, too. This is why not everything that's organic is automatically better for you.

In fact, that bunch of organic celery that you just bought and is sitting in your fridge right now has even more of a natural insecticide, called *psoralen*, within it than its conventionally grown brethren. And that's because, as I mentioned in the last chapter, plants are skilled in the art of deadly botanical poison. Who can blame them, though? There's not much chance they're going to be able to stand up and outrun us anytime soon. So why not try to poison their predators instead?

Yes, and poison us they have. According to USDA estimates that date back more than 2 decades ago, about one-third of cancer deaths in America can be attributed to the diet. Now that part may not shock you, *but many of those cancer deaths are actually attributed to natural carcinogens that are already present in our food.*

Now that's a lot of natural poison!

If you're wondering what natural poisons you may have lurking around your kitchen, let's take a minute for me to tell you more about psoralen. Celery and many of the other plants in its family, such as parsnips, make natural poisons called furocoumarins. Psoralen is one such toxic chemical insecticide that celery uses to fight off insects that want to turn a celery plant into their next meal.

The more that a plant like celery is stressed, say by an insect munching on its leaves, the more psoralen it's going to make in response. In some cases that can mean up to 100 times more psoralen! This means that when celery is grown organically, it can naturally contain much more psoralen than if it's grown through conventional means. The synthetic pesticides used in conventional growing practices kill the insects that increase the amount of psoralen present.

Psoralen has even been known to cause contact dermatitis in workers who harvest it if they're not careful and wearing the appropriate gear to protect their skin. What's really unique about psoralen is that it only becomes "activated" by exposure to sunlight. To avoid being poisoned, some insects have adapted their behavior to counter this bit of chemical warfare and roll themselves up in a leaf to stay in the dark while they happily digest their celery. By doing so, the insects protect themselves from the sunlight that would activate and turn "on" the psoralen that the insects just ingested.

I'm not suggesting that you start only eating celery in the dark, but you need to begin thinking about and become more aware of the natural chemical poisons that may be lurking in your foods, organic or not. This is crucial because natural chemicals like psoralen are not only insecticides but also potent carcinogens. They cause cancer by damaging your DNA—which is why it's really important to be aware of what phytochemicals you eat.

To help protect your DNA and reverse genetic aging, I'll be giving you important instructions on how to avoid or reduce consumption of the phytochemicals that can harm you. It's important that you do this, because once phytochemicals like psoralen become activated, they can target your DNA destructively.

Another vegetable that cranks up its defensive phytochemical arsenal when it's unhappy is the parsnip. Grocery store research sampling has found that old parsnips can contain up to 2,500 percent higher levels of furocoumarins (remember that psoralen is one of these) than fresh parsnips. That's why I've included parsnips in the "Avoid" category of the DNA Restart Health Tip #24, since it's impossible for you to know the amount of furocoumarins present in the parsnips you purchase.

This brings me to a very important takeaway point: *Not all phytochemicals are phytonutrients.* Some phytochemicals can be really helpful to you by, for example, lowering your risk for cardiovascular disease, such as salicylic acid (similar to the main active ingredient in aspirin, acetylsalicylic acid, a precursor of which originally came from the bark of the willow tree). But even too much of a good thing like aspirin can cause you harm. It's all about the dose. A few other examples include theobromine and caffeine, found in cacao and coffee beans respectively. In high enough doses, both of these phytochemicals are neurotoxic to insects. Yet in the right amounts, both can be beneficial to human health.

So how do you avoid getting too much of the DNA-damaging phytochemicals like psoralen?

Do not eat too much of the same food. Your body and the DNA within it actually evolved to have their needs met by having you constantly consume a variety of foods. *Monoeating*, as I've named it, is the single best way to ensure that you're going to be deficient in some nutrients and poisoned by getting too much of others. This sounds like simple advice, but when it's not followed, you can get yourself into trouble quite quickly. A patient of mine decided to go on an organic juice fast, and that, combined with a day at the beach, resulted in severe skin burns. This was because the organic celery juice that was used for her juice fast was full of psoralen, which then became activated while she was on the beach. She was very fortunate and healed without scarring, but it could have been a lot more serious.

Besides advising you not to eat too much of the same foods, I'm going to introduce you now to some foods that you will completely avoid on the DNA Restart. I'll also tell you which foods you need to limit your intake of on your DNA Restart and beyond because they contain natural chemicals that can harm and age your DNA and so shouldn't be consumed excessively.

Cucurbitacins, for example, are highly toxic natural chemicals that can damage your DNA and are found in vegetables such as cucumbers. They also happen to be extremely bitter and can even be tasted at one part per billion, or ppb. So if you happen to bite into a bitter-tasting cucumber, especially if it was organically grown, better to leave it on the plate than risk harming your DNA.

Goitrogens are phytochemicals that are found in vegetables such as turnips. It's thought that they might be responsible for a small number of goiter cases every year because they interfere with your thyroid's natural ability to take up iodine from your

---

 ## DNA Restart Health Tip #12

Here's a list of foods to avoid consuming because they can harm your DNA.

1. Too much of any one food (No monoeating!)
2. Dried fruits (this includes raisins)
3. Peanuts
4. Apple juice and apple cider (nonalcoholic)
5. Commercially produced applesauce and puree
6. Overly bitter cucumbers
7. Large amounts of kale, Brussels sprouts, cauliflower, or mustard greens
8. Raw kale
9. Visibly bruised or moldy fruits and vegetables

diet. So on the DNA Restart you are forbidden from sourcing all of the vegetables in your diet from raw *Brassica* member plants such as kale, Brussels sprouts, cauliflower, and mustard greens. Enjoy these delicious veggies cooked, since many of the dangerous isothiocyanate compounds luckily become inactivated with cooking. So *no more raw kale, because it's full of goitrogens!*

Mycotoxins are another group of naturally occurring chemicals that are produced by fungi that have infected and contaminated one of your food sources. *Aflatoxins* are produced by some strains of the microbe *Aspergillus flavus* and are some of the most carcinogenic and toxic natural compounds known. Besides being able to cause liver cancer, they can also cause birth defects and directly damage your DNA.

The possible presence of aflatoxins (there are four types, with B1 being the most potent and common in foods) is the reason that you are forbidden from eating any peanuts or peanut-containing products during your 28-day DNA Restart. Imported peanuts can often contain much more aflatoxin than peanuts produced in the United States, but there's no way for you to ensure that your peanut butter has been properly tested and is aflatoxin-free. As you'll see in Chapter 17, titled Feed Your DNA Lots of Nuts, there are many other healthy nuts to choose from that don't involve risking damage to your DNA. Aflatoxins (along with other different mycotoxins) are also found in many dried fruits, which is the reason you're not allowed dried fruits during your DNA Restart.

Patulin is another mycotoxin that is produced by different microbes that can often sneak their way into your food. This mycotoxin sometimes finds its way onto your dinner table in apple juice, applesauce, or even apple pies. Often commercial producers use moldy apples to make these products, because the processing hides the fact that they are spoiled and should not be consumed. That's simply disgusting. Patulin in bruised apples is why you're not allowed to consume any commercially produced apple products on the DNA Restart; you should also watch out for things that have been sweetened with apple juice, as you're not allowed those either.

There's one way to drink your apples and not get exposed to patulin, and that's by drinking hard apple cider. Interestingly, when apple juice is fermented into an alcoholic drink from apple juice, patulin breaks down naturally. With the notable exception of an occasional unsweetened excellent-quality hard apple cider from a local microbrewery, there's truly no reason for you or your family to consume apple juice or otherwise processed apple products, since they've been shown to damage your DNA.

Ochratoxin A is another mycotoxin produced by microbes that also damages your DNA. It can make its way into your diet through contaminated foods such as dried fruits, bruised apples, and improperly stored cereal grains. It's also been detected in

R_X DNA Restart Health Tip #13

To protect your DNA from the pesticides, fungicides, insecticides, and herbicides that are used in conventional farming, you should aim to buy organically grown varieties of the following fruits and vegetables, to protect your DNA.

1. Apples
2. Citrus
3. Cucumbers
4. Grapes
5. Peppers

6. Potatoes
7. Stone fruits like nectarines and peaches
8. Strawberries
9. Tomatoes

many milk-based infant formulas, cereal-based baby foods, and apple-based baby foods as well. So avoid all of these on the DNA Restart!

After all this talk about the many toxins naturally present in your food, you may have lost your appetite a little. I've never forgotten the reaction I got from one of my patients when I described how toxic some of these phytochemicals can be.

"So what's the point of eating?" was her response.

After all the gloomy talk of phytochemicals and mycotoxins, I understand why you might be thinking the same thing. But there's a very big bright side, and that's the immense benefit of protecting your DNA from inflammation and oxidative stress that is provided by dietary phytochemicals that are found within phytonutrients, especially those derived from fresh fruits.

Just remember that an apple a day will not keep the doctor away if it's full of mycotoxins. So choose your foods wisely!

MEN NEED MORE CHOLINE THAN WOMEN

Every cell in your body demands choline. It's used as the basis to make neurotransmitters in the brain and to support important structures like cell membranes. It's even involved in helping to transport cholesterol. Choline also plays a pivotal role in protecting your DNA strands from physically breaking apart. This probably explains why studies have found that when groups of people are deficient in choline, they seem to be at higher risk for certain types of cancers.

We can get most of our required choline from our diets, but because this is such an important nutrient, our bodies have the genetic ability to make some as well. But like nearly everything else I've been telling you about, the amount of choline our bodies can make and need is dependent upon the DNA you inherited. Some of us, such as the 60 percent of American men who've inherited a version of the *MTHFD1* gene, need much more than others, in some cases twice as much.

Yet the biggest thing that determines how much choline you need is whether you inherited DNA that turned you into a man or a woman. Because choline is so crucial to fetal development, a woman's DNA is primed to be able to make more of it and is signaled to do so by the presence of the female hormone estrogen.

Having less circulating estrogen, men are at much greater risk of not being able to make enough choline, especially if their DNA isn't great at making their own choline to begin with. In recent years this deficiency has been compounded by the fact that many men have been cutting out of their diets one of the great sources of choline: egg yolks! Ninety-nine percent of the choline in a whole egg is actually located exclusively in the yolk. So as many more men have been opting for egg-white omelets, they've been unwittingly setting themselves up for a dangerous choline deficiency.

Not getting enough choline is a big deal because not having enough can lead to fat accumulating in your liver, a condition called *nonalcoholic fatty liver disease (NAFLD)*.

My wife was bugging me for the last few years to get a physical, because she was noticing that I was tired with low energy, and some days I felt like I was walking around with a foggy brain. I thought I was just stressed out about work, but when I finally went to my doctor and finished with all of the testing she sent me for, she told me that I had nonalcoholic fatty liver disease. I went on the DNA Restart and made all of the changes. I just got my tests back and can't believe that the DNA Restart totally fixed all my fatty liver problems. I was on a high-protein diet for years, which kept my weight down, but I had no idea that throwing away all those egg yolks trashed my liver. Who knew? I had never heard of choline and didn't know that men don't get enough of it. Now I'm telling everyone that food can really be medicine.

—Tom, 57

This condition can eventually put you at risk for cirrhosis of the liver and eventually even liver cancer. NAFLD can often present itself through elevations in liver enzymes, through tests that are often included in routine bloodwork. If that wasn't risky enough already, men without the ability to make enough choline are also at risk for infertility, because choline is required for healthy sperm production.

Given these facts, if you're a man or a postmenopausal woman, you should be making sure that you're getting enough choline from your diet, which for some of you can be as high as 650 milligrams a day. That's why whole eggs are encouraged on the DNA Restart (a large egg can contain 125 milligrams of choline)—especially if you can find ones that are pastured, as they are a great and easy way to ensure you get enough choline for your DNA. Other foods that are rich in choline and acceptable for your DNA Restart include (limited portions of iron-rich) animal products, seafood, nuts, and vegetables such as broccoli.

If you're going to do your DNA Restart as a vegan, you should increase your weekly nut intake, as I advise in the meal plan section later in this book.

YOUR DNA LOVES DIETARY MINERALS

Many of the enzymes (which are like miniature protein machines) that fight oxidative stress, repair your DNA, and reverse aging thrive on metals like copper, manganese, selenium, and zinc.

That's why it's so crucial to get enough minerals from your diet. It's not just in healthy enzyme production that minerals play such a key role, but also in structural skeletal support, like the main role *calcium* plays in your bone health. But minerals like calcium are also used extensively by other systems in the body, including the nervous system in conducting messages. Additionally, calcium plays a role in the clotting cascade that stops you from bleeding when you experience an injury like a cut.

Copper and *manganese* are other important metals we need to get from our diets, with rich sources being cocoa products, nuts, legumes, oysters, and whole cereal grains. If your food is being sourced from nutrient-poor soil, you could be getting a lot less copper than your body needs.

Copper-zinc superoxide dismutase and *manganese superoxide dismutase* are enzymes that help your body mop up free radicals that produce oxidative stress in the body. Getting less than adequate amounts of copper can reduce the amount of this superoxide dismutase enzyme that your DNA can make. Be careful, though, if you're taking supplements, because some may contain too much copper. High levels of copper in the body can do more harm than good by leading to higher levels of oxidative stress. Remember what I mentioned at the start of this pillar about the

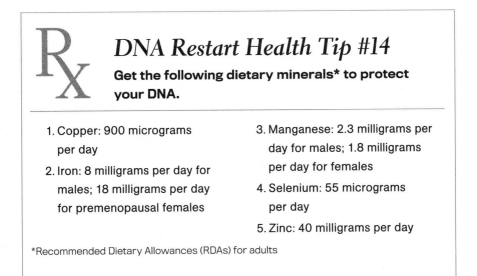

R̲X̲

DNA *Restart Health Tip #14*

Get the following dietary minerals* to protect your DNA.

1. Copper: 900 micrograms per day

2. Iron: 8 milligrams per day for males; 18 milligrams per day for premenopausal females

3. Manganese: 2.3 milligrams per day for males; 1.8 milligrams per day for females

4. Selenium: 55 micrograms per day

5. Zinc: 40 milligrams per day

*Recommended Dietary Allowances (RDAs) for adults

delicate balance between DNA damage and repair? The same principle applies when it comes to supplementing your dietary mineral consumption: Make sure you're getting enough, but be aware that overdoing it can be biologically and genetically detrimental.

Selenium is another trace mineral needed by the body, with Brazil nuts being a particularly good source. Many proteins in the body incorporate selenium in their structure. It's also used to make an enzyme called glutathione peroxidase, which helps to lower the level of oxidative stress in the body. In addition, selenium is thought to assist your cells in repairing their DNA. It's also been shown that a form of selenium called *selenomethionine* can turn on your *TP53* gene, which makes a really important tumor suppressor protein called p53. In some controlled experiments, researchers found that if they stressed cells with UV radiation but gave them a good supply of selenium, they could get a threefold increase in p53 activity, which also doubled the amount of DNA repair that could occur.

This is likely how selenium is involved in lowering the risk of many types of cancers. But before you rush off to start supplementing your diet with extra selenium, be warned that taking too much selenium can lead to toxicity and negative health consequences. Where exactly your food comes from can also have a very big impact on the level of selenium it naturally contains. Certain parts of the United States and Canada have soils that are naturally higher in selenium than other areas in the world, such as Scandinavia and parts of China, that are deficient. This is another reason not to practice monoeating, consuming the same foods produced from the same farms year after year, even if they're local and organic. In the case of selenium, by consistently eating the same foods grown in the same places, you can end up with a deficiency or toxicity, depending upon the amount that's naturally present in the soil where the food was grown.

The human body can contain as much as 3 grams of *zinc*. This essential trace mineral is used by more than 300 enzymes in the body and plays an important role in maintaining a properly functioning immune system. Zinc is also important for the B cells in the pancreas that produce insulin.

The best dietary sources of zinc include fish, legumes, nuts, oysters, poultry, and red meat. Zinc is an example of a mineral that's absorbed better by the body when it's eaten along with a meal containing protein. The reverse can also happen, as less zinc is absorbed when eaten with foods that contain large amounts of fiber and phytate (more on phytates in Chapter 18). So eat your zinc with protein!

CHAPTER 14

The Juicy Step to Reverse
Genetic Aging

Mice do it, cats do it, dogs do it, and even elephants do it. For some unknown reason humans and our primate cousins (and, yes, guinea pigs, too) are the only mammals that cannot make their own vitamin C. While the rest of the mammalian world happily and effortlessly takes glucose and turns it into vitamin C, we have been condemned to get ours from food alone.

If you've ever wondered how your dog or cat survives on dry kibble just fine while guinea pigs need fresh veggies and fruits, now you know. It's the same reason, the *ability* or *inability* to make vitamin C.

It's thought that humans lost the capability to make vitamin C some 40 million years ago, which left us—along with gorillas and chimpanzees—dependent upon fresh food to survive.

We actually still have the same gene used by other animals to make vitamin C from glucose (in humans it's called *GULOP*)—it's just that our version looks genetically like someone cut out parts of it to make a paper napkin snowflake. This means that no matter how hard our DNA and body try, we're not going to be making vitamin C anytime soon. It's also one of the major limitations on our species' ability to travel long distances without a fresh supply from food.

We haven't figured out a way to fix the *GULOP* gene yet, and so until then you are completely dependent on consuming this key vitamin to shelter you from the damage caused by oxidative stress to your body. It's not only oxidative stress that's a concern, though, because vitamin C is an essential cofactor in multiple biological functions in your body, such as maintaining collagen and making important things like hormones and neurotransmitters. It's also water-soluble, so it easily flows through your bloodstream, mopping up nasty pro-oxidant compounds (remember, those guys are DNA damaging).

Here's where things get interesting, though.

Although you lack the gene to make your own supply of vitamin C, that doesn't mean the genes you've inherited don't play a very big role in how much you need.

The individual genetic variability of vitamin C requirements was known long before this vitamin was even discovered. Back in the 18th century, British seafarers who were stuck at sea for months at a time noticed that some of the sailors seemed to be able to cope a lot better than others when it came to long voyages without getting a lot of fresh fruits and vegetables in their diets.

And that's because we all differ in the genetic versions of the genes we have inherited that are involved in things like moving vitamin C from the gut and into the cells and bloodstream. The version of the specific gene you have inherited (such as the genes *SLC23A1* and *SLC23A2*) will determine how much vitamin C gets into your body and cells.

This is one of the reasons we don't all require the same amount of vitamin C from our diets—it all depends on which genes you've inherited from your ancestors. It seems that some of us, like particular British sailors, are much better at getting by without a lot of vitamin C than others. Thanks to advanced research studies, we now know that certain genes some people inherited also make them require more vitamin C because these genes don't work as well to prevent oxidative damage that can then harm their tissue and DNA.

An example of this is the *HP* gene that comes in two main flavors: *HP-1* and *HP-2*.

As you remember from the 1st Pillar, having excess iron freely floating in your bloodstream is never a good idea because it increases oxidative stress in your body. That's where the *HP* gene comes in. Its job is to make a protein, called haptoglobin, that binds hemoglobin (which contains iron) that may be floating around in your bloodstream and, in so doing, reduce oxidative stress. Like a star employee, the *HP-1* gene does an exemplary job, easily outperforming *HP-2*.

Not surprisingly, it's the *HP-2* that's getting continually reprimanded by the body for failing to pick up enough of the free hemoglobin in the blood. Your body has no

---

R̶X *DNA Restart Health Tip #15*

Enjoy the genetically beneficial properties of lemon or lime juice every day.

1. Find one lemon or two small limes.

2. Juice the lemon or limes.

3. Transfer the juice into a 12-ounce glass.

4. Fill the remainder of the glass with sparkling water or still water.

5. Repeat every day for 28 days.

way to fire this gene and is stuck having to mop up all the oxidative stress mess that's left in the *HP-2* gene's wake. The mop in this case is vitamin C, and having more on hand helps to keep the body's elevated oxidative stress levels in check.

Given what I've told you about the increased need for vitamin C, wouldn't you like to know which HP gene you've inherited?

To protect yourself from any potential for genetic discrimination that comes with all genetic testing that's done today, I'd rather you go old school on this one, as there's no reason to submit to costly and chancy testing when you can do one simple thing that's almost free. As it turns out, no matter which version of the *HP* gene you've inherited, *HP-1* or *HP-2*, studies have shown that if you make sure to get enough vitamin C, it doesn't matter if this gene is misbehaving.

And that's why as part of the DNA Restart, you're not going to bother with costly and risky genetic testing. Instead, you're going to make like a sailor and, without exception, have the juice from a lemon or two limes every single day for 28 days.

MAKE LIKE A BRITISH SAILOR EVERY DAY

You may be familiar with why British people are often referred to as limeys. It's because of the amazing discovery centuries ago by early seafarers that drinking lime juice can have seemingly miraculous health benefits.

The 18th-century Scottish physician Gilbert Blane had something incredibly powerful in his lifetime, and that was the ear of the British Admiralty. And Blane's advice was simple. If the Admiralty wanted to have British sailors keep their teeth firmly rooted in their mouths and stop their gums from incessantly bleeding, they would need to start drinking the juice from lemons or limes. Daily.

Blane had no idea why citrus might be having all of these powerful health effects on the sailors who were drinking their juice. And that's because his recommendation predates the discovery of vitamin C.

CHAPTER 15

The Right Fat Combinations for Reversing Genetic Aging

During my junior year in high school, I decided to join the wrestling team. I loved the combination of mental effort and physical stamina that the sport required to truly excel. Due to the rigorous physical demands of the sport, I was also spending a few hours every single day training.

So after a routine physical before the start of the next wrestling season, I was a little more than surprised when my doctor informed me that my blood test results indicated that my cholesterol was elevated. I still remember sitting in his office that day as I was proudly sporting my new wrestling team leather jacket, patches and all. My doctor was quite baffled, and so was I. He asked me a few dietary questions and shrugged his shoulders. He suggested that maybe this was just an inaccurate result as he made the arrangements for me to return for a follow-up appointment, but this time he asked that I fast before the next blood test.

I left that day completely puzzled. I wasn't guzzling raw egg yolks like some of my teammates. In fact, the only real change in my diet was that I was eating more salads than before. Lo and behold, a few days after my follow-up blood test, my doctor's office gave me a call and asked that I come in to get my results.

Now I was really stunned—my LDL, or "bad," cholesterol was really through the roof! And not only that, but my HDL, or "good," cholesterol was lower than average. The only good news was that my triglycerides were within the normal range.

My doctor now sat me down for "the talk" and let me know that given my age and how high my LDL and how low my HDL was, according to the statistics of people in my situation, I was at risk for cardiovascular disease if I didn't make some immediate and serious lifestyle changes.

I asked what he meant by "lifestyle," and he informed me that the three key things to getting my cholesterol right were exercise, weight loss, and a strict low-fat diet.

The funny thing was that as an elite high-school athlete at 16, I definitely wasn't overweight, and as for exercise, I couldn't imagine doing more than the 2 to 3 hours a day I was already doing. As for my doctor's lifestyle changes, the only thing remaining that I could really modify was my diet.

So off I went with my list of foods to avoid. I was told to come back in 2 months after making the necessary dietary changes. These included no longer eating butter and switching to "cholesterol-free" margarine instead. I also was to cut out or minimize all foods that contained cholesterol. This encompassed foods such as cheese, yogurt, eggs yolks, and meat. I was further instructed to remove a few other evil foods, such as coconut and chocolate. The list was long and brutal. Essentially, every food that made eating a pleasure was to be removed.

I discovered in the fine print of my doctor's dietary recommendations that he encouraged me to eat more protein and carbohydrates instead of fats. Okay, that sounded a little more doable. And so it began.

After 2 months on my low-fat diet, it was time for me to go back to my doctor's office to get my blood retested and wait for the results. I felt proud of myself for following my doctor's clinical dietary directives so completely and never veering once in the whole 2 months by having any product containing more than a trace amount of fat. I was certain my hard work would reflect itself in stellar, normalized blood test results. I knew something was wrong, though, as my doctor looked more serious than I was used to seeing him at my follow-up appointment.

He asked me to sit down and then he passed me a printout of my results. I couldn't believe the numbers in front of me. Not only had my LDL cholesterol not decreased after 2 months on a low-fat diet, it actually went up! And if that wasn't bad enough, my HDL, or "good," cholesterol went down.

What was happening here? "So much for low-fat diets" was all I could think. The funny thing was that my doctor was convinced that I wasn't sticking to a low-fat diet and must have been cheating.

What's amazing to me is that it's been many years since my low-fat dietary experimental disaster, and not that much has changed in regard to standard medical recommendations for someone in my situation. In fact, it was only after I myself designed and then completed the DNA Restart that I saw a significant improvement in my own cholesterol profile.

GENES, MUFAS, AND PUFAS

Irrespective of genetic predispositions, there are some fats that are nasty for all of us, and some that are essential building blocks for genetic health and vitality. And most important for people like me, some fats can actually help you keep your cholesterol levels in perfect balance.

Although fats have been vilified for years, if you know the right ones to eat, you can lose abdominal or belly fat, decrease joint pain, lower your triglycerides, and even decrease your risk for breast cancer.

What's important to remember is that all fats are also very energy dense at 9 kilocalories per gram, whereas both carbohydrates and proteins are less so at 4 kilocalories per gram. Proteins require more energy for your body to break down, so they are actually the least energy dense as well as being very good at keeping you feeling full for longer. It's important for you to understand why you need to remove certain fats from your diet, because it's going to be one of the most crucial components to reversing and preventing processes involved in genetic aging.

Because not all fats are created equal, we're going to spend a little time discussing their differences so that you'll make the best dietary choices during your DNA Restart. Some of this information is rather technical, so I'll provide a useful summary of what you need to know when it comes to fats for your DNA Restart at the end of this pillar. Let's get started.

Monounsaturated fatty acids, or MUFAs, are a prime reason why olives are revered for their health benefits. Humans have been eating olives for a really long time. The first olives are thought to have been cultivated more than 6,000 years ago. Olive trees are really long lived and can be upward of 500 years old.

Olive oil is an example of a plant-derived source of fat that's very rich in MUFAs, at around 75 percent. It's a good source of omega-9s, particularly oleic acid (also found in macadamia nuts), which helps lower LDL cholesterol.

Technically speaking, olives are a fruit. And the amount of MUFAs doesn't vary much among the three main grades of olive oil: extra-virgin olive oil, virgin olive oil, and olive oil. But there are very significant differences among them. Extra-virgin olive oil is considered the highest grade, and the lowest grade is simply called olive oil, which is in principle a seed oil, since it's derived from the olive pit.

Only extra-virgin olive oil is derived purely from the flesh of the olive without using any chemicals or heat. Because of that, when a bottle is labeled "virgin" or "olive oil," you are to avoid it completely on your DNA Restart.

Another thing that differs significantly among the grades of olive oil is the amount of phytonutrients from the 230 different compounds that have been identified. These include phenolic compounds, triterpenes, and phytosterols. These phytonutrients are actually found in much higher concentrations within higher grades of olive oil and can lower elevated inflammatory markers that I mentioned earlier (IL-1B and IL-6), which is obviously very good for your genes and overall health.

But the level of phytonutrients can also vary among varieties of olives, where they're grown, and even between seasons from the exact same farm. As olive oil is increasingly processed, the quality of the oil itself decreases along with degrading the important phytonutrients it once contained. That's why it's so important for you to only buy and consume the best-quality extra-virgin olive oil during the 28-day DNA Restart.

Paying a premium for a better olive oil will get you the greatest amount of phytonutrients, such as *oleocanthal* and *hydroxytyrosol*, for each of the 9 kilocalories you're getting for every gram of oil you're eating. To increase the amount of phytonutrients that reverse genetic aging for the same amount of energy or calories, go for only the best-quality extra-virgin olive oil.

It's important to always store all of your oils away from extraneous light and air, so opt for opaque bottles that seal well to make sure your oil doesn't oxidize or become rancid, losing many of its health properties. And remember, paying more for a genetically healthful product is an investment in your genetic health for decades to come. It's so worth it.

Yet it's not only for health reasons that you should be consuming better-quality oils such as extra-virgin olive oils. They also taste so much better and go further. And that's what you want—to combine the most flavor, so that you use less and get the most health benefits all at the same time.

MUFAs are also found in other foods such as certain nuts, which I'll discuss a little later in detail, as well as avocados and certain seed oils.

Recently, many researchers have questioned the benefits of MUFAs. It's now believed that their health benefits depend on their source, as in animal (like beef tallow or lard) versus plant, and how they are eaten (such as in fried food versus cold in salads).

What I believe is causing some of the mixed research results, besides the way the studies are designed, is that all MUFAs have very different phytonutrients within them. In fact, I believe that it's these phytonutrients themselves, like oleocanthal and hydroxytyrosol in extra-virgin olive oil, that are significantly contributing to a healthy body and DNA.

Polyunsaturated fatty acids, or PUFAs, such as alpha-linolenic acid (ALA), docosahexaenoic acid (DHA), and eicosapentaenoic acid (EPA) have a much better track record for improving your potential for genetic health, while other PUFAs, such as arachidonic acid (AA), promote genetic aging largely by increasing inflammation.

Let's explore this a little more.

Your body cannot produce some PUFAs on its own, and these are called essential fatty acids.

PUFAs play a very important role in both disease prevention and progression. Diets that are rich in certain omega-3 PUFAs such as ALA, DHA, and EPA have all been connected with lower incidences of cancer and cardiovascular disease.

Omega-3s and omega-6s are not fixed end products, as your body has the genetics to use complex biochemistry to convert different PUFAs within the same family group because they all have somewhat different functions. An example of this would be linoleic acid (an omega-6 PUFA), which can be turned into arachidonic acid (another omega-6 PUFA) by the body. Linoleic acid was initially thought to be a cause of inflammation that's associated with cardiovascular disease, but that's now being questioned because many of the studies used linoleic acid sourced from trans fat margarine.

There are two intertwined thoughts as to why you should be eating more from the omega-3 group, such as ALA, DHA, and EPA, than from the omega-6 group, such as AA. Both groups are needed, but it's the balance that's important. For starters, omega-6s such as AA lead to more inflammatory products, such as leukotriene B4, which have also been recently shown to promote insulin resistance in mice. Omega-6s can also lead to more platelet aggregation through thromboxane A_2, which can make your blood more prothrombotic, or likely to clot. Thromboxane A_2 is what low-dose aspirin targets to reduce the incidence of heart attacks due to clots.

On the other hand, some of the things the body produces from omega-3s are anti-inflammatory as well as antithrombotic. It's also thought that increased consumption of both EPA and DHA can lead to more resolvins, another group of metabolites that are thought to have potent anti-inflammatory actions.

Even though your body can make DHA and EPA, it doesn't seem to be so great at it, which is why I want you to make sure that you get as much as you can from your diet. The best source of DHA and EPA is often fish, which is why they're often called marine omega-3s. I'll be delving more specifically into which types of fish are particularly recommended on the DNA Restart later on in this pillar, since not all fish contain significant amounts of DHA and EPA. DHA is also really important for neurological development because it is epigenetically involved with at least 100 genes that all govern neurological functioning.

So with all of that information, you might be tempted to focus exclusively on omega-3s and totally eschew omega-6s. But it's important to remember that both omega-3s and omega-6s are needed for your body to function optimally. Unfortunately, because so many of the farmed fish and animals people are consuming today are being fed diets that are high in omega-6s, when we eat them, we end up with an extra dose. That's too much omega-6!

An easy way to move the balance in the omega-3 direction is to use some ground flaxseed or its oil, since it's a great source of ALA as well. On the DNA Restart, you'll be required to get 2 to 3 grams of ALA each day. To do this, on the days when you're not eating any nuts that are rich in ALA, such as walnuts, you'll need to have about a tablespoon of ground flaxseed.

WE DON'T ALL HANDLE FATS IN THE SAME WAY

Think about what happened to me when I followed my doctor's dietary instructions carefully and reduced my fat intake. My LDL cholesterol actually shot up! Just like when we spoke about some people's ability to break down lactose as an adult, the latest genetics research continues to underscore the importance of the role genes play in how we respond to certain diets.

An example: A study published recently in the journal *Science* reported genetic findings that helped explain how the Inuit living near the Arctic Circle thrive on a high-fat diet consisting of foods such as blubber, seal, walrus, and fish. These genetic changes that gave the Inuit this ability resulted in much lower LDL cholesterol than is to be expected. Interestingly, almost all of the Inuit people from Greenland who were studied have this genetic change, but only 2 percent of Europeans and 15 percent of Han Chinese were found to share this genetic similarity. But it's not only cholesterol that was affected, as it seems that the entire profile of fatty acids normally found within the blood is also altered.

What does that mean?

Well, each biochemical conversion step in the pathway of producing and utilizing omega-3s and omega-6s occurs through multiple enzymes that are all genetically encoded. Genes do everything! Some of us are born with enzymes that work better than others when it comes to omega-3 and omega-6 conversion, just like those enzymes that are involved in the breakdown of alcohol, which you discovered firsthand when you did your DNA Restart Cotton Swab Alcohol Intake Test. And this special genetic ability to metabolize omega-3s and omega-6s exceptionally well is how the Inuit and a few rare Europeans and Han Chinese thrive on a diet that's high in fats.

When it comes to the genetics behind the metabolism of fats in the body, a lot of scientific work remains to be done. We're still not sure which exact genetic change has made the Inuit so well adapted to a high-fat diet. But thankfully, there's a lot you can start doing today and much more to come.

For example, using the DNA Restart Cracker Self-Test gives you your optimized carbohydrate-eating plan. Although the Inuit have genes that allow them to eat more fat than most people, if they were to take the DNA Restart Cracker Self-Test, they would likely find themselves within the Restricted category. This is because, as a group, the Inuit probably have a low copy number of the *AMY1* or amylase gene, as researchers found in the Yakut who live nearby and near the Arctic Circle.

So what do the research results from studying the Inuit ultimately teach us for the 2nd Pillar, Reverse Aging? That we need to think about our genetic ancestry and consume only the right fats, in the right amounts, to reduce and not promote inflammation. I'll be delving into this in more prescriptive detail in the next section.

There's nothing worse than chronic inflammation for your overall health, but especially your genetic health. So let's get started with a plan to get you eating fats the right way, genetically speaking. The overall goal is to get you to reduce your intake of saturated fat in your diet and increase your PUFA intake, especially from the omega-3s.

GOOD FATS AND BAD FATS

As you now understand the basic mantra of this book, for genetic reasons we don't all respond in the same way to the foods we eat. Yet there are some food truisms from a genetic perspective that can help us all live a healthier and longer life.

So let's focus on what you can change. Irrespective of your Carb Consumption Category (Full, Moderate, or Restricted) on the DNA Restart, you should never consume more than 40 percent of your daily energy intake from fats. And in the case of the Full Carb Consumption Category, your daily fat intake should never exceed 30 percent. Because you'll be limiting your red meat intake to not more than twice a week (as per the 1st Pillar), it will naturally be much easier to limit the amount of saturated fat you're consuming. Yet keep in mind that oil is like energy-dense fuel; it clocks in at 9 kilocalories per gram, more than double the energy from carbohydrates and proteins. So you should choose your fuel wisely and try to get out of your lipids (a fancy word for fat) as much phytonutrient-sustaining antioxidant and gene healing power as you possibly can. Read on to find out how.

To take you a step further on your journey, I dove deeper into research on oils so that I could prescribe for you the top plant-derived oils to help your genes thrive. All of my recommendations come from fruits and edible seeds and nuts, and ensure the highest level of genetic health.

In the DNA Restart you can cook with and consume only oils from fruits, nuts, and some seeds that are edible in their raw, unprocessed state. So you'll be allowed to have oils that come from olives as well as coconut and other nut sources such as hazelnut, pistachio, and walnut, because you can find and consume these nuts in nature for their oils. And they're packed with powerful phytonutrients that are good for your DNA. You are not allowed to consume either butter or ghee, since they lack the impressive phytonutrient profile of the plant-derived oils that are recommended on the DNA Restart. As well, depending upon the genes you've inherited, your LDL cholesterol may increase significantly with your consumption of both, which is another reason to only use fats and oils from the DNA Restart–approved list. Specifically for cooking, you'll be allowed to use extra-virgin olive oil and coconut oil at lower temperatures than you might be used to.

This means no more canola oil, soybean oil, safflower oil, cottonseed oil, and the like in your cooking or anywhere else. Right now the most consumed oil in America is from soybeans, and there's one big reason for that: It's cheap and easy to market as "cholesterol-free," which is a marketing trick, since all oils from plant sources are naturally cholesterol-free. Further, soybean oil is often partially hydrogenated, making it an even poorer choice for consumption. So let's move on to the good news and talk about some of the oils that are delicious and good for your DNA.

Medium chain triglycerides, or MCTs, have been gaining a lot of popularity recently and are being consumed largely in the form of liquid coconut oil. It's still early scientifically speaking, but some initial data indicates that by replacing long chain triglycerides in the diet with medium chain triglycerides, like a liquid coconut oil, modest weight loss occurs without changing your blood lipid profile significantly. MCTs contain medium chain fatty acids (MCFAs), which makes them an exception when it comes to how your body uses them for energy. MCFAs are taken out of your gut and shunted directly to your hepatocytes, cells found in your liver. Once inside hepatocytes, these fats are oxidized in the mitochondria, which, as you no doubt recall from the start of this pillar, are the energy-processing plants of your cells. So MCTs have the potential of being metabolically converted into fuel for energy production instead of being kicked over and stored in adipose, or fat, tissue. More energy, less fat is good news all around.

You're highly encouraged to use MCT oil in your 28-day DNA Restart (it's one of the approved oils), especially when you're doing low-temperature pan frying or cooking. The one drawback with MCT oil is that it lacks some of the phytonutrients that come with extra-virgin olive oil. Yet in some recipes that have been developed

R℞ DNA Restart Health Tip #16

Keep your DNA safe and ban all of the following fats and oils from your kitchen and your DNA Restart.

1. Butter and ghee	7. Margarine
2. Canola oil	8. Palm oil
3. Corn oil	9. Peanut oil
4. Cottonseed oil	10. Safflower oil
5. Hydrogenated oils	11. Soybean oil
6. Lard	

for the DNA Restart, a much better taste result was achieved in some food creations when MCT was used sparingly, since its unique taste can be mistaken for butter. But don't overdo the MCT and forget olive oil, because its antioxidant and polyphenol profiles are stellar.

HOW TO BALANCE YOUR OMEGA-6 TO OMEGA-3 RATIO FOR YOUR GENETIC HEALTH

Remember how I mentioned that a higher intake of omega-3s has been associated with many positive health outcomes, everything from a reduction of abdominal "belly" fat to dampening of inflammation that leads to genetic aging?

Well, one of the best and easiest ways for you to increase your intake of omega-3s is to have at least two to three servings or more of an acceptable fish or seafood per week (see DNA Restart Health Tip #17 below for acceptable seafood choices). Cold-water fish is also an excellent source of vitamin D. Many people today are still not getting enough of this important vitamin; having darker skin, which blocks the sun that helps the body make it, or spending too much time indoors can lead to serious deficiencies. Having darker skin can put you at increased risk for certain types of cancers when there's insufficient sunlight to meet your vitamin D needs. Increasing your consumption of fatty cold-water fish (that are DNA Restart approved,

DNA Restart Health Tip #17

Have two to three servings (2 to 4 ounces each) per week of seafood from the following prescribed list. And remember to keep your fish selection varied to decrease your exposure to the same contaminants.

1. Atlantic mackerel
2. Butterfish
3. Herring
4. Mussels
5. Oysters

6. Rainbow trout*
7. Sardines
8. Shrimp**
9. Wild king (chinook) salmon
10. Wild sockeye salmon

*Responsibly farmed

**Sourced from the following countries: USA, Canada, Central America

R℣ DNA Restart Health Tip #18

Protect your DNA and *avoid* the following seafood choices that are unacceptably high in methylmercury and other contaminants.

1. Ahi	7. Marlin
2. Albacore tuna	8. Orange roughy
3. Bigeye tuna	9. Shark
4. Chilean sea bass	10. Snapper
5. Grouper	11. Swordfish
6. King mackerel	12. Tilefish

such as wild salmon, Atlantic mackerel, and herring) can help you meet your vitamin D needs.

If eating more fish that are DNA Restart approved is not a dietary option—either because you're choosing to follow the DNA Restart as a vegetarian/vegan or you just don't like fish—then I recommend you get more omega-3 and vitamin D as a daily supplement.

If you're going the vegetarian/vegan route, it's very easy today to find an appropriate high-quality omega-3 supplement, of which I suggest you take at least 500 milligrams per day. If you're open to having a fish source, than you can't go wrong with an intake of 500 milligrams of cold-water fish oil every day. As for vitamin D, you can add a high-quality supplement to give you 1,000 IU a day.

The other thing to keep in mind, to protect your DNA when you're making seafood choices, is the levels of contaminants like methylmercury to which you might be exposing yourself. Some heavy metals make their way naturally into seafood because they're both present in greater amounts in certain environments and accumulate in seafood at different rates. Yet many of the contaminants that we're ingesting today are from human sources like coal-fired power plants.

It's not just pregnant women and children who are at risk—so is your DNA. In DNA Restart Health Tip #18 above, I've provided a list of seafood choices you should avoid; be sure to consult it when making your seafood choices. Seafood offers another example of the dangers of monoeating, as eating the same seafood choices over and over will quickly cause your body to accumulate toxic levels of contaminants. That means don't just stick to one seafood choice.

I'm sure by now you've heard all of the reasons that you shouldn't be eating artificially manufactured trans fatty acids, or "trans fats," so I'm not going to repeat them here. Most of the trans fats you've been eating up until now were likely added to processed foods you might have been consuming. Restricting consumption of all processed foods while on the DNA Restart is required, since processed foods are also full of emulsifiers, so it's going to be a lot easier to avoid trans fats than you might be imagining right now once you rid your pantry of emulsifiers.

CHAPTER 16

Olives and Their Oil for DNA Protection

I think so highly of the genetic health potency contained within the olive that I'm dedicating this chapter to them. Although we touched on olive oil previously when discussing MUFAs in the last chapter, one of the best ways to help keep your DNA free of the ravages of oxidative stress and inflammation that are associated with aging is to consume olive oil.

For the last few decades the heart-healthy oil of choice has been olive oil, but it's only very recently that we have begun to understand that many of the benefits of olive oil are happening on a genetic level.

As we've learned about epigenetics previously, your genes can be turned on and off, or if you imagine them as musical instruments, their volume can be cranked way up or down depending on environmental signals. And those, of course, include the food you eat.

Paraoxonase 1 (PON1) is one such gene that's great to have turned on nice and loud. When the PON1 gene is working harder, it's thought to play a big role in protecting your cardiovascular system from disease by preventing fats in your body from turning "rancid," or becoming oxidized.

Preventing fats within your body from becoming oxidized is important because oxidation can set you up for DNA damage. It is also a trigger for atherosclerosis, or hardening of the arteries. And since cardiovascular disease is the number one killer in the developed world, having more PON1 available to get rid of oxidized fats that play a big role in the process of atherosclerosis is extremely important. But that's not all it does. It also helps your HDL, or "good," cholesterol work harder to help keep your heart healthy. The only problem with PON1 is that as we age, our DNA produces less of it.

So how do you get the PON1 gene working harder for you?

One way to get more positive work out of your PON1[1] is through consuming

---

[1] Not everyone responds in the same way to olive oil consumption because there are many different versions of PON1 that you can inherit. Even if you've inherited a version of PON1 that doesn't respond in the same way to olive oil, that's okay, because there are so many additional positive things that this oil can do for your genes.

extra-virgin olive oil. Extra-virgin olive oil also plays a big role in stopping inflammation and reversing oxidative stress—two of the biggest players in causing your DNA to start aging.

The three main compounds that are found in olive oil are *carotenoids, polyphenols,* and *tocopherols.* Many of these compounds work either to mop up oxidative stress or dial down inflammation. Some can even do both.

In particular, *hydroxytyrosol* and *oleuropein* have been flagged as extremely potent phenolic antioxidants found in extra-virgin olive oil. In addition, another phenolic compound found in extra-virgin olive oil, *oleocanthal,* is getting much more attention within the scientific medical community, as we now know that it can inhibit the cyclooxygenase, or COX, enzymes in the body, just like many pharmaceutical medications can, which serves to dampen inflammation. It also has some neuroprotective effects and has been successful in reducing the inflammatory markers that have been associated with Alzheimer's disease.

The highest grade of olive oil—extra virgin—is the least processed of all, and therefore has also been shown to have the highest level of phenolic bioactive anti-inflammatory compounds and antioxidants present. That's why it makes sense to feed your body only the best grade of oil possible, and that's extra virgin. Research

R̶X̶ DNA Restart Health Tip #19
Decrease inflammation and oxidative stress by eating olives.

1. Eat at least three or four olives twice a week or one serving of the Ancient Antioxidant Olive Tapenade (page 227) once a week.

2. Make sure your olives still have their pits. The process of removing the pits (as well as stuffing them) has been shown to reduce the amount of phytonutrients present in the olive.

3. Try different varieties of olives— they all contain different levels of phytonutrients.

4. Do not eat olives that have gone through a California-style oxidation method. These olives are dark black and produced in California. They've gone through an unhealthy, artificial chemical process that results in a reduction in their antioxidant profile.

backs up extra-virgin olive oil's superior health benefits; it has been shown to have the greatest impact reducing the degree of oxidation to LDL cholesterol as well as reducing other inflammatory biomarkers found in your blood, such as C-reactive protein (CRP). Polyphenols especially found in olive oil are also thought to be behind the reduction in cardiovascular disease that happens when people are ingesting them.

The actual olives themselves and not just the oil can also give you many health benefits, with some studies finding up to 3 percent of an olive's weight coming purely from polyphenols. That may not sound like a lot, but a handful of olives, or even going all out and eating some homemade Ancient Antioxidant Olive Tapenade (you can find the recipe on page 227), can really go far in boosting your anti-inflammatory and antioxidant genetically based reparative ability.

Just one kalamata olive can have about 3 milligrams of hydroxytyrosol, which is as much as you get in an entire 16-ounce bottle of extra-virgin olive oil. In fact, there's data indicating that all you need is a minimum of about 5 milligrams of hydroxytyrosol to start reducing LDL oxidation—that's only two kalamata olives a day!

CHAPTER 17

Feed Your DNA Lots of Nuts

On your DNA Restart, I'd like for you to aim to have up to four servings of nuts a week from the list provided on the opposite page. Eating four servings a week of nuts is a great way to boost your intake of phytochemicals, including antioxidants, which you're going to hear more about shortly and that help protect your DNA from aging. Nuts are also a great source of healthy fats such as MUFAs and PUFAs, minerals such as magnesium and potassium, dietary fiber, protein, essential oils, and antioxidant vitamins such as tocopherols.

Some of the best choices include chestnuts, hazelnuts, pecans, pistachios, and walnuts in particular. I want you to completely stay away from peanuts because of the possible risk of exposure to mycotoxins, as we discussed in Chapter 13. Nuts are also one of the richest sources of antioxidant phenolics (only certain spices and fruits contain more), with chestnuts, pecans, pistachios, and walnuts being especially good sources.

Catechins and *gallocatechins* are abundantly present in hazelnuts, pecans, and pistachios. (They are also found in tea, which you'll be hearing more about in the 4th Pillar, Drink Oolong Tea.) These compounds are good for your DNA because they shift your genome to stop the damaging cascade of transcribing and translating inflammatory genetic signals, which leads to genetic aging. Another group of polyphenols are the *proanthocyanidins* that are abundant in almonds, hazelnuts, pecans, and pistachios.

When it comes to flavonoids—a subclass of phenolic compounds—almonds, hazelnuts, and pistachios are great sources, with pecans reigning as king. *Stilbenes* are another group of health-promoting compounds that are found in pistachios. Pistachio skin also contains resveratrol, a stilbene that's been previously associated with drinking wine.

These are just a few of the thousands of phytochemicals found in nuts. There's profound phytochemical diversity among them all as well. You might be wondering why this impressive phytochemical portfolio is present in nuts.

The reason? Making a tree (which is what nuts were trying to do before we got interested in ingesting them) is complicated business. All of these health-enhancing phytochemicals were originally intended to support the life of the fledgling tree and help it thrive. Many of these compounds are actually found within nut pellicle or skins, so you should try to eat your nuts with their skins still intact, as this will maximize the phytonutrients you are getting without really adding any additional calories. Nuts are also a great source of MUFAs as well as PUFAs—chestnuts and walnuts are especially good sources. However, the oil content of nuts can be high, as in the case of macadamias, which have a 75 percent oil content, or low, such as in chestnuts, which are only around 3 percent oil. This is why even though nuts contain a healthy mix of both MUFAs and PUFAs, they are still all relatively highly dense sources of energy and so should always be eaten according to the specific serving size allowances below. And remember the dangers of monoeating: Since not all nuts are created equal, don't stick to the same handful; be adventurous in your nut of choice.

All of this nut goodness has now been backed by a recent meta-analysis, a type of powerful research study that combines smaller studies to get more accurate results. The results indicate that consuming just 1 ounce of nuts four times a week, for a period of 4 weeks, can reduce your chance of dying from ischemic heart disease by 24 percent. In addition, the researchers found that consuming nuts can reduce your triglycerides, LDL cholesterol, and even your blood pressure. Their data soberly concluded that in 2010, nearly 2.5 million people died from not eating enough nuts.

So on the DNA Restart, you will eat nuts. Not as a between-meals snack, but rather within and at the end of your meals. Snacking is not allowed during your 28-day DNA Restart. The goal is to get you eating and staying full from your

---

R_X *DNA Restart Health Tip #20*

Have up to four servings (1 ounce each) of nuts* a week from the following list.

1. Almonds	6. Macadamia nuts
2. Brazil nuts	7. Pecans
3. Cashews	8. Pistachios
4. Chestnuts	9. Walnuts
5. Hazelnuts	

*Don't forget to vary the nuts you consume so that you can get an array of different phytonutrients.

DNA-optimized Restart meals, three times a day. But there's a much more important reason for eating nuts toward or at the end of your meal that you'll hear more about in "Always Eat Dessert for Your DNA" on page 122.

Finally, I'll end the chapter with another reminder about peanuts: This nut, which is really a legume, should be avoided at all costs. As we discussed earlier, it's not so much the "nut" that I find problematic, but rather the nasty toxins that can seriously harm your DNA and promote the development of certain cancers such as hepatocellular carcinoma. So don't eat them!

Now, let's get back to some more DNA-health-promoting food.

CHAPTER 18

Legumes Are Good for Your DNA

E aten for thousands of years by people all over the world, legumes are not only a great source of fiber and protein—they are packed full of antioxidants, folate, and various minerals.

Legumes have been given a hard time lately for the many types of phytochemicals that are also found within them. Many DNA Restarters have asked me whether they should be eating legumes, given that they've heard that certain phytochemicals present can interfere with their digestion. For example, lectins—a group of phytochemicals found in many legumes—have the potential to interfere with the absorption of nutrients in the human gut.

So why would I possibly be recommending that you eat defensive phytochemicals that don't have your optimal health in mind? Because I like science.

Scientific research has confirmed that the consumption of legumes reduces both cardiovascular and coronary heart disease and even lowers cholesterol.

You shouldn't be eating legumes just for the fiber, though, since other vegetable sources can also provide this. What legumes have to offer is a very unique and rich source of isoflavonoids and phytosterols, which your genes want you to consume.

The question that DNA Restarters raised is not at all new and was even addressed thousands of years ago by the ancient Greeks, who always advised to thoroughly cook legumes before consuming them, no matter how hungry your guests might be!

Here's the lowdown: Indeed, some of the phytochemicals found in legumes are defensive—remember what I told you about plants not wanting to be eaten. That's why presoaking and sprouting legumes are two important ways to naturally reduce many defensive phytonutrients. It's also a fun habit to get the whole family involved in as you tend to your very own kitchen "sprout patch." But if presoaking and/or sprouting is not feasible, there is a DNA Restart–approved, time-saving alternative. There are a few brands of canned beans on the market that both presoak and properly cook their beans, before packaging them in BPA-free containers.

Troublesome phytochemicals can also be handled through proper cooking methods using moist heat, such as boiling. This has been shown to decrease lectins by almost two-hundredfold. Just make sure that you don't ever use a slow cooker to prepare your legumes. The problem with this method of cooking is that temperatures lower than 176°F do not destroy lectins.

I'm excluding soybeans because I believe they should be eaten only after they've gone through a fermentation process—such as in miso, tempeh, or natto—that does an excellent job of removing defensive phytochemicals. More on this in the 3rd Pillar, Eat Umami.

I'm also excluding fava beans from the DNA Restart because there is a genetic condition affecting more than 400 million people worldwide called G6PD deficiency, or favism. Fava beans contain phytochemicals such as vicine and covicine that can be deadly to those with favism.

Chickpeas in particular are rich in both biochanin A and formononetin (much more so than are soybeans), which is important because they activate peroxisome proliferator-activator receptor-α (PPAR$_\alpha$) and β (PPAR$_\beta$). These receptors are involved in helping your body get its lipid or fat profile and sugar balance right, respectively.

Phytosterols are phytochemicals that physically compete with cholesterol absorption in your gut. This is likely one of the ways legumes help reduce the amount of cholesterol that gets into your blood. A meta-analysis of studies that looked at legume consumption found that they can play a role in reducing certain cardiovascular diseases.

As for the phytates that are contained within many legumes, yes, they can bind some of the essential minerals in our diets, making them unavailable. But much more importantly, they also bind heavy metals such as arsenic, cadmium, chromium, and lead. Think of having a little phytate with your meal as having a leguminous botanical dietary Brita water filter that cleans out your gut from unwanted industrial contaminants present in all of our food and environment at large.

I'd rather you lose those heavy metals than gain a little extra minerals. As for minerals, I've designed the DNA Restart to ensure that you get a perfect supply of what you need. And as you might remember from the 1st Pillar, almost everyone in America is getting too much iron today unless you're anemic, so phytate will actually help you in that regard as well.

Eating legumes has even been shown to reduce your risk for diabetes. There is now research that links an increased consumption (having three to four servings a week) of legumes as positively correlated with both weight loss and maintenance of that weight loss (which everyone knows is the hardest part). And if that wasn't reason enough for you to start eating more legumes, here's just one more for you.

In a head-to-head inclusion/exclusion study, 30 obese patients were randomly

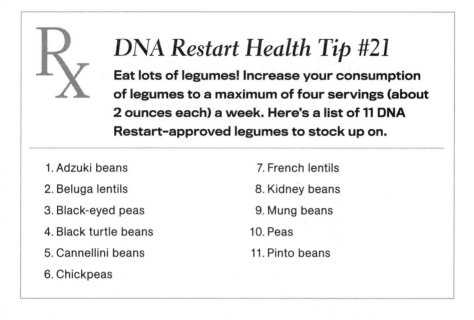

DNA Restart Health Tip #21

Eat lots of legumes! Increase your consumption of legumes to a maximum of four servings (about 2 ounces each) a week. Here's a list of 11 DNA Restart-approved legumes to stock up on.

1. Adzuki beans
2. Beluga lentils
3. Black-eyed peas
4. Black turtle beans
5. Cannellini beans
6. Chickpeas
7. French lentils
8. Kidney beans
9. Mung beans
10. Peas
11. Pinto beans

assigned a weight-loss diet, either including legumes or excluding them. The group eating the legumes had a significant reduction in pro-inflammatory markers like C-reactive protein. These results are promising if your goal is not just to lose weight but to be less inflamed and thus reverse genetic aging.

So sorry . . . but legumes cannot easily be replaced by other vegetables like broccoli.

The overarching goal of the DNA Restart is for you to recognize that there's no one else genetically like you on this planet. So if you're allergic to legumes or something thereof, don't eat them. If you're not, then enjoy and exploit their myriad scientifically substantiated health properties.

CHAPTER 19

The DNA Restart Way
of Food Preparation

One of the most important things you'll be doing in this 2nd Pillar is learning how to prepare foods the DNA Restart way, which will involve crucial cooking techniques that will protect your and your family's DNA from the harms of genetic aging.

Not every beneficial phytochemical gets into your body in the same way. Some are water soluble, like the vitamin B group, and others require fat to be absorbed to be more soluble, such as the carotenoids that give many fruits and vegetables their bright colors.

That's why preparing and cooking foods the DNA Restart way (which I'll be explaining in detail soon) is the crucial key for you to maximize the amount of health-sustaining phytochemicals you're actually getting from your food. The DNA Restart incorporates different food preparation and cooking techniques to always maximize your phytonutrient intake.

Some phytonutrients are normally lost because they can't survive the high temperatures of cooking, while others are leached out of foods, like what happens when you boil vegetables and then discard the cooking water. However, cooking can actually make some phytochemicals, such as capsanthin and carotenoids from red peppers and lycopene from tomatoes, much more available for your body to absorb.

Preparing foods the DNA Restart way will also help improve the nutritional quality of the foods you are eating to maximize their potential health benefits. Most importantly, preparing foods in this way will also ensure that they will be delicious—because delectable and nutritious foods are what your DNA is craving.

This is how you're going to get the most phytonutrients from the food you're going to be preparing during your 28-day DNA Restart.

The most practical way for me to explain how to start incorporating the DNA Restart food preparation techniques is to walk you through a mock-up of a typical meal. The same cooking principles will be applicable to all food preparations. What I'll be demonstrating is a Saffron Chicken with Vegetable Patch Stew and Almonds (see the recipe on page 241).

Let's go back for a minute to the results of your DNA Restart Cracker Self-Test in the 1st Pillar. I've included a table below that lists the three possible Carbohydrate Consumption Categories for your optimal intake, either Full, Moderate, or Restricted. Use your category as a guide for additional carbohydrate allowances for this meal.

If your optimal carbohydrate intake level was Restricted, you don't have to prepare any other side dishes, as you can have the stew as it is. But if your optimal carbohydrate intake level was Moderate, you have the option of eating your stew along with half a portion of brown rice or quinoa. A Full carbohydrate allowance for a dish like this means you can incorporate a full portion of brown rice or quinoa with your meal if you'd like. But remember, regardless of your category, the potatoes in the stew count toward your weekly carb allowance.

When you're finished reading this section, remember that you can turn to page 241 for the complete recipe, including all the measurements and cooking times. So for now, just sit back, relax, and keep reading!

Sample Side Dish Allowance for Your Individual Carbohydrate Consumption Category

Your DNA Restart Cracker Self-Test Results	Carbohydrate Side Dish Allowance
Full	Brown rice or quinoa
Moderate	½ portion of brown rice or quinoa
Restricted	No additional carbohydrates

CHOOSE THE HEALTHIEST COOKING METHOD AND TEMPERATURE FOR YOUR DNA

To begin with, it shouldn't be a surprise that deep-frying is completely forbidden for many reasons on the DNA Restart. This is not just because of the added excess caloric energy from fats but also because the very high temperature potentiates thermal damage and the creation of multiple pro-oxidant compounds like acrylamide that's just not DNA friendly.

The biggest problem with preparing animal protein is that there's usually a soup of pro-inflammatory and DNA-harming chemicals that are produced as a result of the cooking process. *Heterocyclic aromatic amines* are one group of 25 chemicals that result from high-temperature reactions in the presence of proteins. Yes, those delicious scrapings at the bottom of the pan—that's them. Don't be fooled! They're definitely not good for your genetic health. The amount of cooking time, temperature, and cooking techniques all determine how many heterocyclic amines you produce in your dish. Many of these compounds are thought to be carcinogenic when consumed over the long term and definitely harm your DNA. So there will be no grilling or deep-frying of any animal or fish/seafood products during your 28-day DNA Restart. As I'll be explaining further below, it's the high heat produced during cooking that drives the production of heterocyclic amines and *advanced glycation end-products*, or AGEs. When it comes to reversing genetic aging, these two together are your number one enemy. In fact, every time you have ever browned any foods, you've made advanced glycation end-products. But a lot of this damage is avoidable, so stay with me.

To reduce the production of both heterocyclic amines and advanced glycation end-products that harm your body and lead to genetic aging, I'm going to have you prepare your chicken breasts with two very important steps in mind. First, every time you prepare any meat, chicken, or fish, you're going to always marinate it. Overnight, preferably.

So let's prepare a very simple red wine and lemon juice marinade (detailed instructions can be found on page 238), pour it over your chicken breasts, and place them back in the refrigerator. And second, the temperature at which you cook your protein-based dish is crucial for keeping it safe for your DNA. When barbecuing and roasting, it's common to hit temperatures above 400°F, which is great for developing hundreds, if not thousands, of new chemical compounds that are pro-inflammatory and contribute to genetic aging. Simply eating boiled meats every time you tuck into your twice-per-week allowance will not do your palate justice, so that's not the answer either.

Deep-frying just adds unnecessary calories (as you'll soon see with my baked versus fried potato example) and less of a developed, complex flavor when compared to the third and final cooking method I'm going to describe. Deep-frying also cranks up the development of both advanced glycation end-products and heterocyclic amines because of the high temperatures that are often used. That's why the best and healthiest compromise in my opinion is to aim to stew your protein dietary options.

Your food needs to be mouthwateringly delicious and to protect your DNA. So you need to marinate your protein sources every time not just to tenderize your meat, chicken, and fish, but for another important reason as well. Research has shown that marinating with wine prior to cooking can reduce certain types of heterocyclic amines by up to 88 percent. Lemon juice will also help, as lowering the pH, which makes things more acidic, has been shown to reduce the amount of advanced glycation end-products that are produced during cooking. Protein dishes that are stewed rarely go beyond 212°F, and that along with the marinade goes furthest to create a healthy and delicious eating experience.

CHOOSING THE MOST HEALTH-PROMOTING INGREDIENTS FOR THE DNA RESTART

Garlic and onions together have an amazing number of potential health benefits, so both will be used to make the stew. As for the onion, go ahead and dice it and set it aside. But unlike in most other recipes, I'll be doing things a little bit differently here with the garlic. Since the phytochemical in garlic called *alliin* needs to be converted by an enzyme called *alliinase* into *allicin*, which then further breaks down into a handful of other phytonutrients, it's always better to crush the garlic cloves before cooking them. Now that the garlic is crushed, and to make sure that most of the allicin is further broken down into *ajoene, diallyl sulfides,* and *vinyldithiins,* give it 5 minutes to just sit before throwing it into the cooking process. Many of these compounds have incredible health benefits, including protecting your DNA from genetic aging.

But it's not just your DNA that can benefit. Ajoene from garlic, for example, is thought to be at least as potent as aspirin as an antithrombotic agent, fancy speak for something that inhibits blood clotting. Allicin is also antibacterial and antifungal and has been shown to be able to kill *Helicobacter pylori*, a bacteria associated with increased risk of gastric cancer. Studies have also found an association between

℞ **DNA Restart Health Tip #22**
Crush or cut your garlic and wait at least 5 minutes before using it.

To get the most phytochemicals out of your garlic, you need the help of an enzyme called alliinase. This enzyme is naturally found in garlic, but for it to become activated, you need to damage your garlic. So smash, chop, or cut your garlic cloves! Heat inactivates alliinase, so wait at least 5 minutes before using crushed garlic in your cooking to maximize the amount of phytonutrients like allicin.

garlic and onion consumption and reduced gastric cancer risk. There's also evidence that a high intake of both garlic and onion can reduce the risk of other cancers in other sites of the body, such as the ovaries, endometrium, oral cavity, esophagus, and others.

Interestingly, eating raw garlic can reduce your risk for lung cancer, and it's thought to do this by a very unique method called garlic breath. True story! I'm sure you've experienced firsthand or been the recipient of the volatile oils from raw garlic that are excreted from the lungs after it's ingested. But it's thought that because the lungs are bathed in raw garlic oil, there's a reduction in lung cancer risk. Given all this, it's no wonder that the 17th-century English herbalist Nicholas Culpeper called garlic "a remedy for all diseases and hurts."

Leave the garlic for 5 minutes to let alliinase do its work and convert more alliin to allicin. The reason for the wait is not only to get more allicin, but also because alliinase is heat sensitive and cooking destroys it. So the longer you wait, the more chance alliinase has of maximizing your allicin quotient in the garlic.

Now's a good time to get out some gorgeous saffron. An incredible spice full of very potent phytonutrients, saffron has been used for thousands of years to impart a totally original flavor to food. Crocin, a phytochemical contained within the stigma, imparts a striking crimson-yellow color that has historically been used to beautify fabrics throughout the world. Legend has it that Cleopatra used saffron as part of her very extensive beauty routine.

DNA Restart Health Tip #23

Have more spices! By increasing your dietary spice and herb intake, you'll be protecting your DNA from inflammation and oxidative stress. Here's a list to get you started. Aim to have at least one of the spices on this list every single day.

1. Allspice
2. Ceylon cinnamon
3. Chile peppers
4. Clove
5. Cumin

6. Ginger
7. Nutmeg
8. Saffron
9. Turmeric

R X **DNA Restart Health Tip #24**
Avoid the following fruits and vegetables for the next 28 days.

1. Conventionally grown apples
2. Celeriac
3. Conventionally grown kale
4. Parsnips
5. Conventionally grown strawberries

6. Both organic and conventionally grown soybeans unless they are fermented. Tamari, tempeh, miso, and natto are exceptions to this rule.

To make the spice saffron, thin threadlike stigmas from the flowers of the saffron crocus (*Crocus sativus*) are separated from each flower. An incredible 75,000 saffron crocus flowers must have their stigmas handpicked to produce only a pound of saffron! It's not surprising that it's one of the most expensive spices in the world. Thankfully, very little can go a long way to impart a delicious and distinct flavor to your food.

As you might have picked up by now, though, it's not just delicious that the DNA Restart is after. Everything that you eat and do over the next 28 days needs to be good for your DNA as well.

Well, both crocin and crocetin, two phytonutrient carotenoids found in saffron, have been shown to have antitumor and antioxidant effects in both laboratory and human studies respectively. Additionally, a meta-analysis that pooled data from initial clinical trials found that saffron might improve the symptoms and the effects of depression, premenstrual syndrome, sexual dysfunction and infertility, and excessive snacking behaviors.

Although further scientific work is still required to build upon these findings, there's no reason to wait before adding some saffron to your life. So do that now. Buy saffron only from a reputable source. This could be a local grocer whom you trust, a reputable spice shop in your city, etc. Given the price that saffron can fetch, there's always the chance that what you're buying is not really saffron. Avoid the powdered form and only buy saffron that comes in a package that allows you to visually inspect the stigmas. It is expensive, but you need only about 0.01 ounce for this dish.

Now presoak the saffron threads in a very small bowl or cup with boiling water. You'll be adding this to the stew later with the vegetables, but for now you can put it aside and leave it to soak.

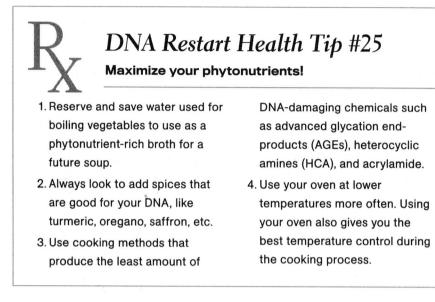

R_X

DNA Restart Health Tip #25
Maximize your phytonutrients!

1. Reserve and save water used for boiling vegetables to use as a phytonutrient-rich broth for a future soup.

2. Always look to add spices that are good for your DNA, like turmeric, oregano, saffron, etc.

3. Use cooking methods that produce the least amount of

DNA-damaging chemicals such as advanced glycation end-products (AGEs), heterocyclic amines (HCA), and acrylamide.

4. Use your oven at lower temperatures more often. Using your oven also gives you the best temperature control during the cooking process.

Of course, saffron isn't the only spice or herb that's good for your DNA. I've given you a list of nine of the most potent spices; try to use at least one of them every day.

Let's move on to preparing potatoes, since many of the same principles apply. When buying your potatoes, let your eyes guide you, since the more color the flesh has, the more phytonutrients and carotenes they'll be adding to your diet. Look for varieties with deeply colored flesh, such as the yellow Yukon Gold or the aptly named Purple Majesty. Whenever possible, choose potatoes that are smaller in size, organically grown, and, most importantly (more on this soon), without any signs of sprouting or turning green. And remember to always have some type of dietary fat included within your potato dish, because the phytonutrients contained within potatoes are fat soluble, so that means their genetically beneficial potency gets unlocked with fat.

Now there are many misconceptions about potatoes—that they're unhealthy and full of empty calories. Let's look at the numbers, shall we?

A 4.5-ounce serving of boiled or baked potato has only about 120 kilocalories. Turn the same amount of spuds into fries by deep-frying, and you can easily hit upward of 400 kilocalories. That's more than a 200 percent caloric increase! I think you get the picture and understand why deep-frying is completely out for your DNA Restart. If you really need more convincing that potatoes deserve your time and a rightful place on your dinner plate, reread the section in Chapter 12 where I detail my visit to the Altiplano in Peru and the calorically balanced, phytonutrient value of the spuds that grow there.

So what I recommend for potatoes is boiling or low-temperature stewing them as the best means of preparation. In this way you're avoiding most of the acrylamide, a toxic by-product of deep-frying or burning, that makes French fries undesirable for your DNA health. It's also why I recommend not burning your toast! Acrylamide can start to form significantly when starch-based foods are cooked at temperatures exceeding 250°F. This is just another reason you're going to be stewing this dish and not high-temperature baking or frying.

But whichever way you choose, boiling or low-temperature baking, you'll need to leave the skin intact. If you don't peel your potatoes before boiling, you'll get 20 percent more vitamin C. The peels provide a useful barrier to the leaching of phytonutrients into the water. So the only potato prepping you'll be doing is to scrub any dirt off them.

Another useful DNA Restart technique when boiling potatoes (or any vegetables, for that matter) is to reserve and save the boiling water. Whatever you do, don't pour that water down the drain! It's chock-full of phytonutrients, so use your boiling water as the base for a future soup stock. Wait until the cooking water has cooled and then refrigerate it, and you'll have a phytonutrient-rich starting point for a delicious future soup.

Okay, now when preparing your potatoes, be vigilant and watch out for those pesky defensive phytochemicals. It's important that you remove or discard any parts of the potatoes that may have turned green or sprouted. This is a sign that they've gone ahead and become mature and begun to produce chlorophyll. Unlike sprouting legumes, when potatoes start sprouting, it's a sign that they have revved up the amount of defensive phytochemicals called glycoalkaloids, such as α-*solanine* and α-*chaconine*, by up to sevenfold, which means you should discard them.

Now remove the green stems from your red peppers but don't deseed them,

---

R̲X̲ *DNA Restart Health Tip #26*
Dining tips that are good for your DNA.

1. According to your DNA Restart Cotton Swab Alcohol Intake Test results, you have the option to drink wine with your meals, but not between them.

2. Always marinate your meat, chicken, and fish to reduce the formation of AGEs and HCAs.

3. Have dessert after dinner—fruits and/or nuts will keep postmeal oxidative stress under control.

unless you suffer from a condition like diverticulosis. It's a great idea to get more red peppers into your diet, especially when you cook them yourself. They are full of capsanthin, a phytonutrient belonging to the carotenoid family, that's often found in cooked peppers at the same level that lycopene is usually found in cooked tomatoes.

Making stews and soups is actually one of the best and easiest food preparation methods to boost the availability of the fat-soluble carotenoids. Multiple studies have found that if you stew your food, you can boost the availability of carotenoids by a third. This cooking method works well because many of the phytonutrients that would typically be leached and lost through steaming or boiling are actually retained within the dish. This also works to improve the flavor profile because many of these compounds provide important components of taste. Stewing also helps to keep the cooking temperature lower than shallow frying or even baking in the oven because of the higher water content.

Always Eat Dessert for Your DNA

You may be surprised to hear that after every dinner, I'm going to encourage you to have dessert throughout your entire 28-day DNA Restart. You can eat dessert as long as it falls into the following categories: fruits, a small square of dark chocolate, and/or nuts. I'm not recommending this because I'm trying to be easy on you. The reason is that whenever you eat a meal, there's always the potential for an increase in oxidative stress and pro-inflammatory processes.

The level of oxidative stress that your body goes through as you're finishing up your meal is related to what you ate. A meal consisting of a serving of red meat, for example, will have more iron, which, as you remember from the section on iron in the previous pillar, is pro-inflammatory to the gut and body. Higher iron intake has even been associated with an increased risk for developing Type 2 diabetes.

It's not just red meat that will do this. The natural transitory increase in sugar and fats in your blood that happens as you're wrapping up eating is also a culprit. Pretty much anything you've just eaten can increase oxidative stress, even if only slightly. So why not help your DNA out and prevent genetic aging?

Eating fruits, a small square of dark chocolate, and/or nuts for dessert at dinnertime is the perfect prescription for decreasing oxidative stress

Now go ahead and chop the rest of the vegetables: potatoes, along with the carrots that you've peeled, and roughly dice your tomatoes and put them aside. Chopping carrots and other vegetables that contain phytonutrients like carotenoids helps to release these compounds and make them more available for absorption into your body.

Now you're ready to start the cooking process. First I recommend adding a table-spoon of water to your pan and then a little extra-virgin olive oil. That helps to keep the cooking temperature down. When the pan is hot, go ahead and quickly sear your chicken breasts on both sides, which have until now been marinating in the fridge. Now add the diced tomatoes, saffron in its soaking water, chopped potatoes, chopped red peppers and carrots, and crushed garlic and chopped onion. Add to your stew 1 tablespoon of double-concentrated tomato paste and 1 cup of water and bring the contents to a boil. There's no need to add salt while cooking. It's better to taste the

and inflammation because it can help increase your body's level of antiox-idants just when you need them most. Studies have shown that by having phytonutrient-rich fruits after your meal, you actually increase the antioxidant capacity of the blood. Everyone knows by now that high-quality varieties of dark chocolate contain health-promoting compounds such as catechins and proanthocyanidins. This makes 1 ounce of dark chocolate (at least 72 percent cacao) an excellent end-of-the-meal dessert for your DNA. To help avoid monoeating, enjoy consuming your 1-ounce serving of dark chocolate as an alternative to fruits and nuts—not more than twice a week.

So now that you're finished eating your meal, it's the best time to have a serving of fruits and/or nuts, which will start mopping up all of the increased oxidative stress and protect your DNA. Remember that the more colorful combination of fruits you have the better, and don't forget to make sure you're eating those nuts with their skins still on (refer back to DNA Restart Health Tip #20 on page 109 for nut suggestions).

And if you're wondering about alcohol, red wine that's rich with antioxidant polyphenols works best if it's consumed during your meal instead of between mealtimes. Just remember to follow your DNA Restart alcohol intake guide from your self-test result in the 1st Pillar, Eat for Your Genes.

And most importantly, bon appétit!

dish as the flavor develops and only add a finishing salt prior to serving. Once it's boiled for 10 minutes, reduce the heat to medium-low and cook, covered, for an hour. Just prior to serving, add a teaspoon of chopped almonds (with their skins intact!) to each portion (remember, this will count as one of your four nut servings per week). Adding cold extra-virgin olive oil is also a great way of getting the phytonutrients from a "cold" oil, so go ahead and drizzle about half a teaspoon of oil over every serving.

Following the steps explained in this chapter is a great way to make sure that the highest amount of beneficial phytonutrients is maintained. Now eat and let your DNA enjoy the nourishing, inflammation-quenching, and genetic antiaging properties of your wonderful creation.

THE DNA RESTART TAKEAWAYS
2nd Pillar: Reverse Aging

1. Activate your innate DNA healing mechanism.

    Find a form of resistance exercise you enjoy and do it two or three times a week.

    Find a high-intensity exercise you enjoy and do it two or three times a week.

2. Fill your shopping cart with fruits and vegetables.

3. Enjoy phytonutrient-dense potatoes in moderation, according to your Carb Consumption Category.

    They're a great way to protect your body from chronic diseases and DNA aging.

4. Avoid harmful phytochemicals that age your DNA.

    See Chapter 13 for details.

5. Avoid all greenhouse-grown fruits and vegetables.

6. Have a maximum of four servings of legumes a week.

7. Ban all mycotoxins.

    This means no peanuts or apple juice, applesauce, or apple puree.

8. Avoid monoeating.

    Don't eat too much of the same foods, no matter what.

9. Remember that men and postmenopausal women need more choline.

10. Eat no more than seven to eight whole eggs (preferably pastured) a week.

11. Your DNA loves dietary minerals, so make sure you're getting enough. This includes copper, manganese, selenium, and zinc.

12. Have up to four 1-ounce servings of nuts a week.

13. Make like a British sailor.
    Drink the juice of a lemon or two limes every day.

14. Use the DNA Restart way of food preparation as a guide when preparing your meals throughout your 28 days.

15. Cook with and consume only those oils from nuts, fruits, and some seeds that are edible in their raw, unprocessed state.

16. End your dinners with one serving of fruits and/or nuts (remember no monoeating, so switch it up twice a week with 1 ounce of dark chocolate with at least 72 percent cacao).

PART III

THE DNA RESTART
3rd Pillar

Eat Umami

No one knew more about how difficult it was to get and stay full than Katherine. Before becoming a DNA Restarter, she'd always struggle with feeling constantly hungry. This led her to incessantly snack between meals. No matter how calorically substantial a meal would be, there was always just something missing for her—namely, that nice feeling of fullness and satiety. This is why not long after she ate a meal, you'd be guaranteed to find Katherine grazing on leftovers in the kitchen or preparing a bowl of granola cereal. This went on for years until she successfully finished the DNA Restart.

How did Katherine bring her body back in line with how we're all meant to be eating? Simple: by eating more umami. As you're about to discover for yourself in this pillar, Katherine found that umami "unlocked" the feeling of fullness that she was ravenously seeking in all those bowls of granola.

UMAMI = SATIETY BOMB

As you're going to see shortly, umami is the real linchpin in the DNA Restart that will get your diet sharply realigned with your genes. Umami is going to be crucial for you to lose those extra pounds and maintain your ideal weight during the next 28 days and beyond, because you're going to learn to feel much more satisfied naturally, with a lot less food, both during a meal and well after you've finished eating it.

For most of human history, our ancestors had to make do with very little in the way of easily accessible nutrition. Famine, not feast, was the reality for most of our genetic ancestors. It's only recently that we've reversed this biological reality, making us, collectively in the developed world, the largest humans in history.

127

Thanks to an incredible feat of human ingenuity in the last century, we've launched an unprecedented botanical technological revolution that has immeasurably increased our agricultural yield. Because of the cornucopia of food today with year-round availability and with such easy access to calories, we live in a world that's completely out of touch with the genes we've inherited.

Our ancestors would be astounded to see the sheer volume of production outputs from factory farming, the flood of milk per cow, the size of our fruits and vegetables, and the amount of animal products at our relatively "cheap" disposal. But most importantly, they would also probably think most of what we produce tastes plain horrible.

To make matters worse, more than any other time in human history, today you're likely to be surrounded by food that's effortless to obtain. And unfortunately, most of this easy-access food is nutritionally junk.

Umami is one of the most potent ways to guide us back to increased feelings of satiety. It uniquely does this by imparting flavors that linger long and strong in our mouths—way after we're done eating. I used to hear repeatedly from many of my patients that their biggest challenge in getting to, and staying at, an ideal weight was that they never felt full and weren't able to even imagine holding on to that feeling. So they quietly snacked in-between meals, raided their fridges hungrily at 11:00 p.m. every night, and routinely felt guilty the next morning. And each and every one of them started melting off the weight after I told them what I'm about to disclose to you: There is another way. A better way. And the key is umami.

What we all need is umami at each and every meal, and it's the key to getting us back onto a healthy track of eating and living.

In this pillar I will teach you the secrets of umami that I gleaned directly from some of the top chefs around the world, such as Chef Nobu Matsuhisa, of the famed Nobu Restaurant chains; celebrated Peruvian chef Flavio Solorzano; and chefs at Michelin three-star restaurants like Daniel Humm, Corey Lee, Yoshihiro Murata, and David Kinch.

In Japanese, *umami* means "delicious," and in this pillar I will show you how to eat the right types of umami foods in just the right ways so that you will lose the weight and actually feel satiated. Called the fifth taste, umami signals our bodies through genetic means that the food we are consuming is abundantly nutritious and full of specific amino acids that are essential for a healthy life.

CHAPTER 20

A Quick Tour of the Five Basic Tastes

Our ability to taste food is actually genetically encoded in all of our DNA to make sure we stay healthy. Each one of our five accepted basic tastes plays a genetic role to protect us from consuming anything harmful and encourage us to eat more of the foods that inspire optimal physical health.

So let's take a quick tour of the five basic tastes. We'll start with most people's favorite: sweetness. Why is it that we find sweet such a delectable taste, and why is it so hard to resist? That's because sweet almost always indicates that what we're consuming is safe and energy rich. For our genetic ancestors, sweet meant food that they could eat and not be poisoned. The reason that we don't have a genetically encoded "off" button when it comes to eating sweet-tasting foods (such as, for example, the way we do with poisonous bitter foods) is because these foods were just plain rare in the environment all of our genetic ancestors lived in. And even when they were present, they were a ton of work to procure. So in biological terms, sweet = precious, rare, and safe to eat.

In all of the circumstances in which our ancestors lived, most of the sweet treats were very seasonal and few and far between. If you could revisit the wild ancestors of all our domesticated fruit trees, you would be surprised to find that the origins of our modern supersize apples, cherries, strawberries, and blueberries are all *mini-me* versions of what we see in the supermarket today.

Plants want to invest only the minimum amount of energy that's required, and that's usually in the form of sweet-tasting fructose, to convince you to eat their fruit and, in so doing, spread their precious seed. So in essence what food manufacturers have done is hot-wire this natural adaptation by adding an immense amount of different types of sugars to a multitude of products, which then signals your body, "Eat me! I'm sweet so I'm safe, and you never know when you might find something like me again." For most of human history, having access to this amount of sugar was a complete ecological impossibility. In fact, any of your genetic ancestors who would be naturally averse to sweet-tasting food would have been less likely to survive. Eating the rare purely sweet-tasting foods that were available helped our various genetic ancestors to survive the rest of the time, given the scarcity of year-round, easily attainable fruits and vegetables.

Grab a saltshaker, as we're now on to the second basic taste. A salty taste can be triggered by the presence of electrolytes such as sodium and potassium in the foods we consume. Electrolytes and minerals are crucial for you to keep your metabolic and nervous systems functioning optimally.

Unlike eating sugar, if you consume too much of a salt like sodium chloride, you can easily and quickly die. In fact, we have a built-in protection mechanism in place to keep this from happening by having an innate taste aversion to consuming too much salt. Think about the last time you spit out a spoonful of overly salty broth. That's kudos to your latent metabolic biological DNA salt-level checker. As our levels of sodium consumption climb, we're neurologically reminded to cut back our salt intake, and signals of thirst bubble up into our consciousness, which are all orchestrated to bring down our levels of circulating salt. This finely tuned system helps to automatically monitor our blood to make sure that we don't acutely overdose.

Time to talk about sour. Think about biting into a lemon wedge. Now you're into sour. In small doses, some of us deeply enjoy the taste of sour. As a matter of fact, when I was speaking with Chef Daniel Humm, whose Michelin three-star restaurant Eleven Madison Park was also ranked number five on San Pellegrino's World's 50 Best Restaurants in 2015, he called sour "the essential component to any successful dish . . . it wakes people up, gets their attention. . . ."

But if sour is very dominant in the food we're eating, it can be interpreted by our

Salty-Sweet Visualization Exercise

Wish you could trick your body to have the same innate aversion to sweets that it has to overly salty foods? Try this quick DNA Restart salty-sweet visualization that DNA Restarters who came before you swear by: If you just can't stop yourself from eating one more doughnut, imagine sprinkling a tablespoon of coarse iodized salt (I'm not talking about that delicious, flaky fleur de sel) on it and imagine what it would be like to take bite after bite, swallow after swallow of that "pinch your tongue" supersaltiness. Do this visualization when the urge to consume the sweet stuff comes up, and you'll have some serious problems just thinking about having a taste.

body as a warning signal, like our aversion to salt, but in this case it's that the food we're eating might be spoiled. The reason for this is that many of the microbes that are involved in food spoilage change the chemical composition of the food, making it taste a pronounced sour, as they work their way through it. Just picture taking a swig of soured or spoiled milk and you'll get the picture.

On to bitter. Our ability to taste bitter even in trace amounts is one of the most important genetically based mechanisms humans inherited to keep us safe. As you may remember from the 2nd Pillar, Reverse Aging, many of the poisons that are naturally present in our foods are from a chemical family called alkaloids. Most plants do not appreciate being roasted or turned into salads. To keep hungry people and animals away, they produce thousands of bitter-tasting compounds collectively called antifeedants. The name, of course, is self-explanatory.

Plants don't want to be eaten unless, as we've talked about earlier, you're eating their sweet-tasting fruits and, in so doing, spreading their seeds. Many of these chemical alkaloid compounds can poison our metabolic system and predispose us to many different types of cancer. This is why an overwhelmingly bitter taste is usually enough of a signal in and of itself for us to immediately stop eating.

In an ironic twist of fate for our plant friends, however, some of these alkaloids that originally served to protect plants from being eaten by us are now the exact reason that we cultivate and consume them. Everything from nicotine and cocaine to atropine, digoxin, and opium alkaloids were all originally part of a botanical defensive armory, which we now exploit to the max for our consuming pleasure.

Plants are also not the only organisms on Earth that are adept at chemical warfare. As you might remember from the 2nd Pillar, microbes and fungi also love to poison potential predators, and many people are surprised to hear that our food supply can be routinely contaminated by "natural" carcinogens such as aflatoxins that are produced by strains of the *Aspergillus* fungi.

This can happen when common crops such as corn, rice, and wheat are stored in a high-moisture environment that allows for fungal growth. Not surprisingly, most mycotoxins can impart a bitter taste—which is again nature's way of politely telling us to please stay far away.

You may have started hearing a little more about umami recently, as it has been gaining more traction and garnering more attention in the Western world of culinary arts. The Japanese scientist Kikunae Ikeda, who is credited with describing the scientific underpinnings of "essential deliciousness," only coined the term *umami* in 1908. But as it turns out, many cultures have been relying upon the taste of umami to make food more palatable for thousands of years.

Umami is not a "new" taste at all. While the genetics behind our ability to detect umami have only recently been elucidated (in the year 2000), as you'll soon see, almost every cuisine from every part of the world has relied on umami to make things taste incomparably delicious. Since our bodies evolved our exact genes over millions of years to naturally keep us healthy and physiologically balanced, why then from a genetics perspective do we taste umami, and why is it important?

THE GENETICS OF UMAMI

The genetic reason behind our ability to taste umami, and the reason it triggers satiety, is that umami signals to our bodies that the food we are consuming contains important proteins that have been shattered into their basic amino acid building blocks. This makes the umami-rich food that we're consuming very valuable from a biological perspective. Umami likely became much more important for our genetic progenitors after the harnessing of fire for cooking and the introduction of preparation and storage of food through fermentation techniques. This is because both cooking and fermentation can increase umami in foods by releasing more amino acids and nucleotides, which also consequently increases their nutritional value.

Umami's ability to relay the taste of present amino acids such as glutamate and then communicate this nutritional powerhouse to our brains as "deliciousness" then altered our food-seeking behavior to direct us to find foods that are more likely to be protein rich. This umami-inspired food seeking provided our evolving species with good sources of the essential amino acids histidine, isoleucine, leucine, lysine, methionine, phenylalanine, threonine, tryptophan, and valine. According to my research, you can thank umami for the big and productive brains that humanity takes for granted today as the current dominators of the planetary food chain.

Umami, or "deliciousness," is mainly triggered by glutamate and to a lesser degree aspartate. These are both amino acids found in proteins that are released after it has undergone some type of preparation process. The 5'-ribonucleotides inosinate and guanylate are also major triggers for this taste. Being able to sense their presence— remember amino acids are the building blocks of proteins—and having that taste register as delicious creates a positive reinforcement loop that has the potential to alter our food-seeking behavior. It's the amino acids that link together to form proteins that our bodies are hankering after. Some of these amino acids are considered "essential"— meaning we cannot produce them on our own and need to get them from our diets.

So can you guess what I was amazed to discover from visiting highly acclaimed restaurants across five continents and then speaking with their chefs about their coveted secret cooking techniques? Let's assume you guessed correctly. For many of the world's top culinary maestros, the "secret" that made some of their dishes so outstanding and memorable was that they were laced with just the right amount of umami.

UMAMI IS HARD TO DESCRIBE, IMPOSSIBLE TO FORGET

Interestingly, for all its flavor-boosting power, umami is harder to taste on its own. Part of the reason for this is that in most Western cuisines, umami is merely one of a kaleidoscope of flavors that compose the taste of a dish. This is one of the reasons that have made conversations about this important taste and the corresponding language that we use rather challenging to nail down accurately. Because unlike the first four basic tastes that we've talked about, umami is much more ephemeral and, to the untrained palate, harder to describe. Part of the reason for this is that unlike salty, sweet, bitter, and sour, umami is far from a binary sensation and much more challenging to put into exact words. Think about the smartest, most complex, and most interesting person you've ever met. How easy is it to describe that person in one crisp sentence? Umami's like the flavor version of that person. This *je ne c'est quoi* also limited umami *until now* from being used as a strategy to increase feelings of satiety and satisfaction from our food.

As we spoke about earlier, for most of human history, our various genetic ancestors struggled with a lack of food. That's nowhere more apparent than in the ubiquitous creation and use of soup by almost every culture that's been studied. Long before Ikeda's scientific discovery, chefs around the world were using the principles behind the taste of umami not only to improve the palatability of food but also to extend what was then a meager supply just a little further. Put simply, umami is the taste that gives food its delicious kick. That's the reason why soup stock tends to be rich in umami, since it serves to mask the fact that what you may be eating is mostly a whole lot of water.

Proteins from animal and plant products don't have much umami taste in their raw, "natural" state until they are broken down into their amino acid components. This is one of the reasons why the taste of many animal products improves with aging as well as, of course, with cooking as more and more amino acids become liberated, deliciously increasing the umami quotient of food.

One thing's for certain. Dial down umami and we can eat to no end. When it comes to our sense of taste, umami is the crucial genetic legacy that we've inherited from all of our disparate genetic ancestors, and it is one of the essential pillars in the DNA Restart.

It is this exact genetically encoded positive reinforcement that our bodies evolved to keep us effortlessly on the right nutritional track. And if used properly throughout the 28 days of your DNA Restart, umami will be the most delicious key to losing weight and keeping it off.

CHAPTER 21

Umami with Breakfast = Weight Loss

The DNA Restart is going to show you how to exploit a higher use of umami-rich protein every day to help you to lose weight and feel fuller. To do this, you need to make sure that you start your day with a breakfast that's packed full of umami. Research indicates that more than half of young people today are routinely skipping their breakfasts, and for the ones who aren't, they are often consuming a breakfast full of flat-tasting, simple carbohydrates and little umami.

I know just how hard it can be to find the time to have a proper, umami-packed breakfast. During medical training I often hurriedly rushed in the morning to see my patients, and a proper breakfast was the last thing on my mind. But skipping breakfast is such a bad idea from a physiological perspective because it signals your body to tell your genes that you're facing a day full of starvation. Our recent genetic ancestors unfortunately had to face days exactly like these more than just occasionally. And for them, there was no guarantee that a mediocre cafeteria lunch later would rescue them from acute hunger pangs. So their bodies wisely packed away whatever calories were eventually delivered and learned to store those caloric treasures, known today colloquially as love handles.

Recent research published in the *International Journal of Obesity* by researchers from the University of Missouri-Columbia focused on teens who were in the habit of skipping breakfast at least five times a week. Sound familiar? What the researchers did was split up the teens into three groups and follow them for 12 weeks.

One group of teens ate a high-protein breakfast meal, rich in umami, which included eggs, dairy, and lean pork and contained 35 grams of protein. Another group of teens ate a normal-protein breakfast, which consisted of a bowl of milk and cereal that had 13 grams of protein. And finally the third group continued to skip breakfast as they had been doing prior to the study.

According to Heather Leidy, PhD, assistant professor in the department of nutrition and exercise physiology at the University of Missouri School of Medicine, "The group of teens who ate high-protein breakfasts reduced their daily food intake by 400 calories and lost body fat mass." Her research also found that the group of teens who continued to skip breakfast gained more body fat at the end of the 12-week study.

Now why would eating more, not less, during breakfast result in weight loss? Eating more to lose weight may sound counterintuitive, but it's not when you reconsider what I was telling you earlier about your various genetic ancestors' recurring experiences of famine. Not eating breakfast is signaling famine, which primes your body to demand and then safely pack away even more calories for later. Convince your body and, most importantly, the genes passed on by your ancestors that there's no need to panic—by eating a signature DNA Restart wholesome, umami-packed, high-protein breakfast early in the day—and you'll voluntarily be primed to consume less for the rest of the day. An umami-rich breakfast is like a public service announcement to your genes proclaiming that all of its essential amino acid needs have been sated, and there is no need to be on high alert (aka binge eat) for the rest of the day. This means you'll lose the weight without the struggle.

Why fight millions of years of genetic evolution? I can assure you that you're not going to win. So all your days need to start with an umami-packed protein hit. Whenever you have protein, you're unknowingly consuming umami, but you can indeed have umami that has little protein. When it comes to breakfast, though, I'm requiring that you go for both.

The good news is that the less effort a major change takes to make in our lives, the more likely we'll be to keep maintaining it in the long term. This is what makes the Eat More Umami pillar so radical and important. Research consistently warns that any lifestyle change that takes too much effort will just not stick. And if it's not going to last, it's not worth doing. That's because most people who fail to stick with major dietary changes end up fatter than when they started.

And that's where umami is biologically incredible. It taps into your very own genes to help you find and use the one genetically based dietary trick that kept your genetic ancestors' appetite in check.

As an added benefit, Dr. Leidy and her colleagues found that consuming an umami-packed, high-protein breakfast also resulted in a more stable level of glucose during the day. It's clearly not just any breakfast that you need to start having. You need to make sure that you're starting your day eating more umami and protein. We are genetically encoded to eat for umami's sake, and this crucial flavor is the driving force behind feeling nourished and satisfied after a meal.

So if you want to lose weight and reduce unwanted feelings of hunger, make sure that you eat breakfast and that it's packed full of umami-rich proteinaceous foods. Stay tuned for a recipe idea on how to do this on page 223.

UMAMI: BEYOND BREAKFAST

It's not just breakfast that needs to have a big umami punch. You should be making every meal infused with lots of natural umami taste. Have you ever found yourself mindlessly snacking on junk food like pretzels? Something odd and annoying happens when you eat carbohydrate-rich foods that lack umami. You never get full.

However, if you consistently swap your carb-rich, umami-poor pretzel snacks for umami-saturated meals, you'll begin to find it impossible to consider taking up empty snacking again. It's just too dissatisfying compared to how good and full umami meals make you feel. I cannot stress how important it is to have umami present in every one of our meals. It's what's going to help you lose weight and feel satiated for longer.

EVEN BABIES TASTE UMAMI

Our first exposure to umami actually occurs in the womb. Unbeknownst to most people, amniotic fluid is a rich, souplike source of umami. Since all fetuses practice swallowing in utero, our earliest taste experience is umami rich. Newborns are also primed early for and rewarded with lots of umami taste, as human breast milk is a rich source of . . . you guessed it: umami. Human breast milk is actually packed full of glutamate, which triggers us to experience the taste of umami. Interestingly, human breast milk has much more umami, with 15 times more glutamate than cow's milk, and most of the commercially available brands of baby formula are umami inferior, too. And that's why I believe that breast-fed babies are much better at regulating their food intake compared to bottle-fed babies. The umami richness of the breast milk is like a natural signal to the babies' genes that their nutritional needs have been met, so they learn to naturally know when to stop feeding.

UMAMI AND THE ROMAN EMPIRE

Curiously, even ancient Roman soldiers' palates were sustained by umami. *Garum*, a syrupy condiment dripping with umami, was made for them through the fermentation of the intestines and flesh of small fish like Mediterranean anchovies.

As I spent countless hours poring over ancient texts looking for nutritional and flavor trends in the classical world, the one surprise that I had was the sheer number of recipes that actually contained garum. Often herbs were added during the fermentation process to create unique flavor profiles, with every producer having a secret recipe. But regardless of taste variability, apparently there was something everyone could agree upon. Garum really stank.

To mitigate for this, laws were even put in place 2,000 years ago that stipulated how far garum producers had to be outside of a town. I'm not sure I want to imagine the combination of the Mediterranean sun, a hot summer's day, and the pungent odors produced from rotting fish.

But amazingly, with only a few drops, garum could completely change the taste of a dish—making things go from so-so to stunningly delicious.

It was the supreme Roman condiment.

Those Roman soldiers knew what they were doing. Infusing their food with umami wasn't only making the food more palatable, it was keeping them satiated for longer.

Unlike the pricing of some of today's commercially produced condiments, garum was far from inexpensive in the ancient world. During Roman times garum was obscenely priced; a prized jug (approximately 100 ounces) of garum could fetch as much as $15,000 in today's US dollars. Many modern European treasure hunters' hopes of finding amphorae filled with gold coins in Mediterranean shipwrecks were often dashed as they only found countless containers of garum. Why the craze for garum, you may be wondering?

Because umami is the stuff of life—and a little bit of garum goes a long way. Remember what I was telling you about umami's ability to make dishes taste delicious? Well, it also has a powerful ability to make you feel satiated and full. After all, if you're a Roman soldier out for conquest, the last thing you'd want on the battlefield is to have a rumbling stomach.

CHAPTER 22

Umami Brings an
Award-Winning Chef Home

When it comes to umami in Kaiseki cuisine—a type of Japanese multicourse meal usually inspired by the seasons—Chef Yoshihiro Murata is considered a treasured master. I traveled to Tokyo and met Chef Murata at his acclaimed Michelin two-star restaurant Kikunoi. Walking through the small garden, I had no idea that Murata hadn't spent his entire illustrious culinary career in Japan, nor did I know what a pivotal role umami played in shaping his career trajectory.

Before sitting down to dinner, I was taken upstairs to a private tatami room so that I could have the opportunity to find out from Chef Murata firsthand the secret to his world-renowned cuisine.

Seated at a low table, he got up and greeted me. I was wondering what it was like to be a third-generation chef and owner of a family business that now includes a veritable necklace of Michelin stars totaling eight—his two Kyoto establishments were given three stars, and his Tokyo outpost two stars.

Over tea poured into beautiful hand-thrown ceramic cups that epitomized *wabi sabi*—beauty through imperfection—Murata told me his family's story. Beginning in the 17th century, Murata's ancestors were tea ceremony practitioners at Kyoto's famed Kodaiji temple, attendants to the wife of the retainer Hideyoshi. Things were going splendidly for his family until 1868, when they abruptly found themselves out of work during the Meiji Restoration.

The political turmoil and lack of employment meant that Murata's relatives needed to be creative. But there was one thing they knew how to do, and that was to flawlessly execute a killer *Kaiseki*—this is a very specialized type of Japanese cuisine often exemplified by a multicourse presentation of up to 25 dishes that try to highlight the flavors of the seasons and, of course, umami.

But as I discovered from Chef Murata, he didn't easily follow into the family business. While still in college, he stunned his father when he announced that instead of following in his father's footsteps and taking over the family's restaurant, he wanted to cook French cuisine. In doing so, Yoshihiro Murata was diverging from generations of Muratas who came before him. And I'm not sure if you've heard this, but family, tradition, honor . . . a very big deal in Japan.

Packing his bags, Murata spent the next 6 months working his way through many notable European kitchens. But something was missing for the young aspiring chef, and he started to feel restless, with a surprising yearning to return home. I asked him what it was that he couldn't find enough of in European cuisine. He answered in a single word: "umami."

There just wasn't enough umami for his liking, so he decided to head back to Japan and pursue training as a chef of Japanese cuisine. Today, besides running the family's three restaurants in Japan, Chef Murata is also a highly respected published author of a series of cookbooks that highlight his skills as a master of Kaiseki cuisine. And all of this astounding success after his homecoming was thanks to umami. He just couldn't live without it.

But I came to Japan to hear firsthand from Chef Murata what he believes is the best way for people to use the taste of umami to lose weight and start eating healthier.

"You should tell them to drink 2 liters [67 ounces] of *dashi* every day," he said. Dashi is a special type of fish-based broth that is prepared from the shavings of a dried fish, related to tuna, combined with konbu, a type of seaweed. When it comes to pure umami flavor, there's really nothing that can come close to dashi, and it's what gives Japanese cuisine its unique flavor profile. As I was going to discover later that evening, Chef Murata flavors almost all of his dishes with his unique dashi.

But I couldn't help smiling to myself as I quickly totaled the number of hours of work, incredibly hard-to-source ingredients, and culinary mastery required to make 67 ounces a day of traditional Japanese dashi. And then I decided that this is just not a reasonable prescription as a rule in the DNA Restart. So I tried again.

"What about foods that people can eat that are a little less difficult to prepare?" I asked. Knowing how particular Murata is regarding the ingredients that go into his dashi—he even has water from his family's well, almost 300 miles away in Kyoto, trucked to Tokyo twice a week to make sure he gets the flavor just right—I thought perhaps there might be something else he could suggest.

"Yes, it might be hard to get the right ingredients to make dashi . . . well, how about tomatoes?"

Chef Murata was right. Tomatoes are hugely central to the cultivation and utilization of umami throughout your DNA Restart. So in honor of Chef Murata's recommendation, I designed an umami taste test experience using tomatoes.

THE UMAMI TASTE TEST EXPERIENCE

Before I explain how you can eat more umami, you need to do the experiential taste test exercise. On the DNA Restart, your goal is to shed whatever excess pounds you're unnecessarily carrying around. To do that, you need to know exactly what umami tastes like so that you can start recognizing how to use it instinctively to lose weight. I've devised an exercise that will upgrade your tasting ability so that you can recognize umami's unique taste. You can do the umami taste test alone or together with your friends and family.

So let's get started!

You'll need a bite-size slice of a tomato, or even a cherry tomato will do. Make sure that the tomato you are going to use is not too watery or hydroponically grown.

You also need a timer and a way to take some tasting notes—use the table provided in this book or a separate notebook or an electronic device, so that you can easily record what you're experiencing during the umami taste test.

After you take your first bite of tomato, make an umami-tasting note at every 10-second interval for the first 30 seconds—that means you should have three umami-tasting notes at the 30-second mark. After the first 30 seconds you're free to swallow your tomato; now make a tasting note at every additional 30-second interval, until you reach 180 seconds, or 3 minutes in total. Then you're done and we're ready to compare our tasting notes.

How do you take an umami-tasting note? Write down anything that comes to mind, but keep it brief. Also, think about what's happening in your mouth as you chew.

The DNA Restart
Umami Taste Test Experience

Place a bite-size piece of tomato straight into your mouth. Then start chewing and record your first taste sensation at 10 seconds. Remember to continue to write down what you're tasting at every 10-second interval for the first 30 seconds. That means that you should have three separate umami-tasting notes for your first 30 seconds of chewing. Now swallow, but keep timing! Now take another tasting note after an additional 30 seconds, and at every 30-second interval after that up to a total of 180 seconds. Three minutes after you started chewing, record your last tasting note. Use a timer to keep it official.

And don't forget to pay attention to your saliva! As you chew, does the inside of your mouth feel wet or dry? Is more or less saliva being produced? You're going to be hearing a lot about saliva, especially in this pillar, as you're about to experience first-hand the fact that umami has an intimate physiological relationship to it. Don't forget to watch for sensations such as textures and mouthfeel.

At the end of the Umami Taste Test Experience, you'll be comparing your tasting notes with mine.

Let's begin!

The DNA Restart Umami Taste Test Experience

Time Intervals	Tasting Notes (Be sure to include whatever comes to mind.)
10 seconds	
20 seconds	
30 seconds Now swallow.	
60 seconds	
90 seconds	
120 seconds	
150 seconds	
180 seconds	

Now let's review. What did you taste?

Likely the first thing you noticed at 10 seconds was a distinctive tomato taste with the beginnings of sour and sweet notes. At 20 seconds that taste would be combined with some sweetness and/or tart or acidic notes. At 30 seconds you probably noticed that by now there was a lot of saliva produced in your mouth, which is why you were encouraged to swallow at this point.

Let's talk a little about umami and saliva. One of the interesting physiological properties of umami is that it has the ability to trigger salivary and gustatory secretions. The reason is that your saliva is packed with enzymes that begin the digestion process. Having umami trigger the release of more saliva can be a really useful property for many people who are occasionally troubled by having an excessively dry mouth (a condition known as *xerostomia*). Generally as we age, we naturally produce less saliva in response to eating, and research has now shown that you can use umami to help get your salivary juices flowing.

Using umami to jump-start our salivation reflex is important because it also starts the digestion process before we actually swallow. As you remember from the 1st Pillar, Eat for Your Genes, the DNA we inherited from our ancestors plays a big role in how we digest and break down our foods. In that pillar I showed you how testing your own saliva at home by doing the DNA Restart Cracker Self-Test and then using the results was an important weight-loss tool.

Returning now to your Umami Taste Test Experience results, what did you experience after you swallowed the tomato? As you can see from my results, at 60 seconds the tomato taste started to fade even though there was a lingering sweetness. And even at 90 seconds, a full minute after I swallowed, my mouth was still watering.

My DNA Restart Umami Taste Test Experience

Time Intervals	Tasting Notes
10 seconds	Pure tomato taste, acidic, tart, some saliva.
20 seconds	Some sour notes, sweetness developing, much more saliva.
30 seconds **Now swallow.**	Sweetness dominates, much less acidic.
60 seconds	Overall tomato taste is fading, but lingering sweetness, increasingly less acidic. Mouth is still watering.
90 seconds	Tomato taste is almost completely gone; mouth still feels wet from continual saliva production.
120 seconds	Savory taste sensation on the surface of my tongue beginning to form.
150 seconds	Savory taste sensation is spreading, becoming more intense.
180 seconds	Much more intense savory or meaty taste; sensation has spread all over my mouth.

I'm the guy who could outeat anyone at a dinner party. There were some days when I felt like my stomach was bottomless, especially when I was stressed. I always struggled trying to keep my appetite in check. That's probably the reason I could never stick to any diet. Umami totally transformed my life. I found out that by mixing and matching a few umami foods, I could actually feel and stay full for the first time. I myself was shocked when I couldn't finish everything on my plate. The DNA Restart helped me lose 24 pounds and never left me feeling hungry.

— **Matthew, 49**

The really big change in my tasting experience occurred at the 120-second, or 2-minute, mark. With the tomato taste almost completely dissipated, I started to sense a savory and rich taste on the surface of my tongue that was hard to put into words. At the 150-second mark, this savory taste started to spread and seemed to move to other parts of my mouth as well. My last tasting note at 180 seconds was the most interesting. What I experienced was that the savory taste had completely spread and become established throughout my mouth—with a surprising, almost distinctive meaty taste.

How did your tasting notes compare to mine? Once you start comparing your results with others', you will notice their experience might be slightly different than yours but that overall there will be certain overlapping trends.

What I want you to focus on is that lingering savory taste that is maintained long after all the other tastes have dissipated.

Welcome to the wonderful world of umami. Consider your taste upgraded.

CHAPTER 23

Umami Is Universal

Many of the recipes, as well as my thinking about diverse preparation methods that I've incorporated into this book, were shaped and inspired by my travels across five continents exploring cuisines such as Acadian, Peruvian, Japanese, Adriatic, Thai, Persian, northern Scandinavian, Hakka, Provençal, and those from the Cycladic Islands. My goal was to scientifically distill and translate what makes a dish delicious.

To help in those matters, I also visited some of the world's best and most celebrated chefs to learn the gastronomic secrets they use and how umami is employed in their kitchens to make their meals so satiating and memorable.

I visited Chef Flavio Solorzano in Peru at his restaurant, El Señorio de Sulco. Chef Solorzano is known as Peru's King of Quinoa and is a longtime fan of *ceviche*, a dish popular in many of the coastal regions of South America. Ceviche is often prepared with raw fish that is cured in lime or lemon juice and combined with salt and aji chile peppers.

With ruffled hair and the physique of a college shot-putter, his appearance wasn't what I was expecting. He looked more like a nightclub bouncer—one who could more easily snap an unruly clubber in half, with one hand, than delicately and lovingly prepare some of his meatless signature quinoa dishes.

What I learned from Solorzano was that unlike most of his culinary peers, he's not a big fan of using just-caught fresh fish for his ceviche.

This sounded almost heretical to me.

"I've found that the freshest fish doesn't actually make the best ceviche," he went on to explain. "And I discovered this completely by accident. I was actually preparing ceviche for what was going to be a live television show—to a panel of tasters. But to my horror, my suppliers sent me the wrong box of fish, one that wasn't as fresh, and it was too late for me to do anything about it," he recounted to me one morning over coffee at his restaurant.

"So I just went ahead and made ceviche like I always did . . . and the tasting panel, they loved it! It was bursting with umami and was probably one of the best batches I have ever made."

The science behind what Solorzano experienced that day is that when most types of fish age, just like ham or beef, more umami-triggering compounds become available as they are released from the flesh.

It turns out that "aging" fish is a much more common practice than most people realize. Many sushi establishments, especially the more exclusive or higher-end restaurants, also "age" their fish before serving it, looking to capitalize on the natural increase of more delicious umami flavor.

In fact, as it turns out, most really respectable sushi restaurants intentionally do not serve the freshest fish. Solorzano was really on to something.

In a recent interview Jiro Ono, the world's oldest chef at a Michelin three-starred restaurant, who was made famous in the documentary *Jiro Dreams of Sushi*, described how he ages his fish, which he serves raw at his restaurant.

For example, his *maguro*, or tuna, is aged for 3 days before he serves it to his customers. Could this be the reason why President Barack Obama, after having his fill of sushi at Jiro's restaurant, declared that it was by far the best and freshest sushi he's ever eaten? Do you think Obama knew his sushi wasn't actually that "fresh"?

The real lesson here is that time begets more umami. This got me thinking about many of the other common misconceptions when we think about the "proper" ways to prepare the best-tasting food.

But to really dive deep into the science and culinary practice of umami, it was time that I left Peru and returned to Japan. After all, Japan, besides being the country that christened the fifth taste, has to be one of the most food-obsessed countries in the world, with a plethora of world-renowned restaurants and chefs to choose from. Tokyo alone boasts more Michelin-starred restaurants than any other city in the world.

What I wanted to do was discover the secret techniques Japanese chefs employ to harness umami to the max. Ones that I could easily share with DNA Restarters, so that you, too, could start showcasing more umami in every dish you eat.

After I arrived in Japan, my first stop was in Tokyo, where I had the pleasure of meeting Chef Yoshihiro Murata (whose story of umami's magnetic ability to bring him back home from France I wrote about earlier in this pillar) and Chef Nobu Matsuhisa. I first tried Nobu's creations while I was staying on the Hawaiian island of Lanai. With 22 restaurants worldwide and a cell phone that never stops ringing, Nobu is one seriously busy chef.

It was obvious upon meeting him that Nobu industriously manages all of the commercial aspects of running his multiple restaurants worldwide with dedicated passion. But there's one thing that I quickly learned Nobu deeply loves. And that's food. I was curious about what got him started cooking and what were some of his earliest memories he had connected to food.

"When I think about my earliest memories around food, they all have something to do with my mother. . . . You know, the sounds of food being prepared by her in the kitchen . . . like the sound her knife made when she was chopping vegetables, and the gas being lit. It was through eating her food, which she always worked so hard to prepare, that I constantly felt her love. That's what I learned from her, to always put my heart and love in every dish I create and make. Because love tastes good! Just like my food." He then flashed me one of his signature boyish smiles. And yes, his smile is contagious.

One thing's for sure: That love is mirrored in an incredible, almost cultlike celebrity following with quotes like that of actress Kate Winslet, which likened his food to "heaven on earth, and sex on a plate."

As the evening progressed, I was interested to hear more of Nobu's thoughts about the importance of umami in making great-tasting food.

"I always make sure to start with umami and then balance it with the other four basic tastes—it's fun to see the faces of my guests when they taste a new dish for the first time . . . especially when they're amazed by the combination of surprising flavors of what I've created."

"Why do you think people love your food in so many cities around the world?" I asked.

"Of course it starts with using the freshest vegetable ingredients I can find, no matter which city my restaurant is located in. My rule is fresh produce is best. That's the real secret to great-tasting food. But I never forget about umami! That's what I usually start thinking about when I'm creating a new dish. It's funny it was only in the last few years that I used this term. Before that, saying something was delicious was enough, but I really think it's by instinct; you can't make great-tasting Japanese food without having umami present."

My dinner kept being interrupted by new dishes Nobu kept on ordering and wanting me to try—no matter how much I was trying to protest that I was already really full from his umami-laden food.

"I understand that you believe it's important for us to be thinking about umami when we're in the business of creating great-tasting food. But what about satiety?" I asked.

"I think delicious food with umami can leave you feeling satisfied, so maybe you don't keep on eating," he replied. As he said this to me, he winked and slipped another piece of his signature black cod with miso on my plate.

"Any tips on how DNA Restarters should add more umami to their meals?" I asked.

"Yes, look at what you're eating," he said while using this as an opportunity to point to the piece of cod on my plate. "Miso is full of umami, so it makes things taste delicious. For this dish it makes the flavor so much more savory than it would be without it. I would recommend trying to add miso to some dishes. It's a simple way to add umami without a lot of calories. Adding a teaspoon into a pot of pureed carrot soup can go a long way to making it taste fuller, more savory, and richer."

At this point I wasn't sure if he was really trying to kill me with his food or making a joke, but while I was taking some notes about using miso, he proceeded to slip a signature Nobu slider on my plate.

"Dessert!" he exclaimed.

YOU NEED TO EAT UMAMI EVERY DAY

Now that you've had your umami taste upgraded, you're beginning to isolate what umami actually tastes like. You're now going to start consciously increasing the amount of umami you consume at every meal, every day. To make it simple for you, I've created a synergistic umami table (see page 161). I've filled it to the brim with some of the most umami-rich foods known to man, so that you can mix and match these ingredients every day of your DNA Restart and start feeling more satisfied with more taste and fewer calories at each and every meal.

The goal is to start to include at least one umami-rich food every day for the first week of the DNA Restart. Then, starting with the second week, you're going to have at least two umami-rich foods every day. By the third week, you should be up to three different types of umami foods a day. It's actually a little harder than it sounds, but every DNA Restarter before you will tell you it's worth it. Once you start noticing how full and satiated you feel after every meal, and how the pounds start melting off, you'll want to have more and more umami!

I was actually surprised at how powerful umami was in stopping people from overeating. Many people who are on the DNA Restart have told me, as well as what I've come to experience myself, that *as you start to shift to increase the amount of umami you're eating at every meal, you're actually able to eat less in one sitting.*

What makes the pillar of Eat Umami important in the DNA Restart is that the focus of eating is to experience delicious and umami-rich foods. As you'll see from the foods I've listed in the table, not every food will have a significant amount of protein. And that's because certain plants (think tomatoes) have made themselves delicious, especially when they want to be eaten, by increasing the umami quotient.

As I mentioned earlier, there's always going to be umami in cooked protein, but there's not always protein in umami. And on the DNA Restart, you're going to be enjoying both.

CHAPTER 24

The Umami Potential of Plants and Fungi

With a few exceptions, most plants are not rich in umami. That's part of the reason why a lot of plants don't impart a savory or rich taste to a dish.

Tomatoes, as you've just experienced after doing the Umami Taste Test Experience (on page 140), are a colossal exception to this general rule. This is likely why tomatoes have become one of the most popular ingredients around the world in so many diverse dishes and can help explain their ubiquitous presence in almost every cuisine. Tomatoes have truly become global.

Why do ripe tomatoes have so much umami? Let's call it a form of botanical laziness on the part of the tomato plant. Since tomato plants can't just get up and walk over to a new or better growing site, they're relying on you to have their seeds consumed and, "ahem," deposited elsewhere. Although we often speak of them as vegetables, tomatoes are actually a seed-containing fruit of the tomato plant, so from a botanical perspective, they're not vegetables at all. Many other plants use the same strategy when they create edible fruits that are full of seeds. In this way the seeds hitch a free ride to what will hopefully be a better new growing location for their offspring.

So why don't green tomatoes contain just as much umami, you might be wondering? The reason unripe or green tomatoes have much less umami is that they don't contain a lot of glutamate. This was the amino acid that triggered your mouth to water and resulted in that lingering umami you detected during the Umami Taste Test Experience.

It's interesting to note that the amount of glutamate in a tomato, and therefore its umami, is correlated with the degree of its ripeness. The reason that ripe tomatoes are such a great delivery system for a burst of umami flavor is that they naturally contain high amounts of the amino acid glutamate. Without the added glutamate, green tomatoes actually taste much more acidic, kind of like a tomatillo. On a tomato's way to a glorious ruby red, the amount of glutamate can increase by as much as 10 times.

What the tomato plant is doing by increasing the amount of glutamate present as its tomatoes ripen is to time the ripeness of its fruit with the development of its seeds. The ruby red of a tomato not only means it's delicious, but also that the seeds are ready for shipping. Eating a tomato at the peak of ripeness also means that there are a lot more phytonutrients present, such as the carotenoid lycopene. Research indicates that vine-ripened tomatoes can contain more nutrients than those that have been gas ripened artificially. Further, some varietals of tomatoes have more umami than others—bottled San Marzano and Santorini are two examples that are particularly enjoyable for their umami potency. While pricey, these tomatoes go a long way, and on the DNA Restart you'll be spending less on meat (as you'll recall from the 1st Pillar, the maximum allowance is no more than two servings of meat per week) and more on ingredients that pack more umami punch per calorie. So choose your tomatoes carefully; the extra umami is worth the extra cost.

Sun-drying tomatoes or turning them into paste is also a great way to increase the amount of glutamate present per weight. This is why on the DNA Restart, you're going to start adding tomatoes, sun-dried tomatoes, and even tomato paste to your dishes. Adding them to your cooking can go a long way toward increasing the savory or umami quotient present. I've included recipes on pages 226 and 236 that highlight the power of tomatoes to kick up the level of umami and help you feel full.

But before you jump into the kitchen to get started, there's one more DNA Restart technique that will boost your cooking's umami quotient.

STOP DESEEDING YOUR TOMATOES

For years I removed tomato seeds during the cooking process and carefully strained all of my tomatoes when I was preparing a sauce. I picked up this habit from a French chef trainee whom I was living with as a college student. What I didn't know was that I was actually removing most of the umami along with those seeds.

If you've been deseeding your tomatoes but you're looking to eat more umami, you should stop. The inner part of the tomato, especially the gel-like substance surrounding the seeds, is actually the richest source of umami. When you deseed your tomatoes, you're literally throwing out umami!

EAT MORE MISO, NATTO, AND TAMARI

Miso and the soybeans that it's made from can be stellar sources of umami. The Asian tradition of using fermented soybean products to enhance the taste of food and increase its umami dates back at least 2,000 years.

Despite edamame's current popularity, eating it is not going to get you much umami. This is because what triggers umami in soybeans is bound up in its proteins and not readily available to push your umami taste buttons when consumed in edamame form.

That's why on the DNA Restart you're going to be eating more fermented soybean products, like *miso, natto,* and *tamari.* As you'll come to see in my list of umami-friendly foods, fermentation unlocks foods' immense hidden umami potential.

Have you ever wondered why tofu tastes so bland on its own? Its entire umami flavor potential is still trapped deep inside its jiggling white hulk. So switch out your tofu for tempeh, which is made through a fermentation process that will give your genes the umami they need to communicate satiety to your brain.

Filled with protein, phytonutrients, minerals, and, most importantly, umami, miso is a fermented soybean paste that's well worth stocking in your kitchen. Miso is so bountifully full of umami, it was the one condiment that Nobu gave a special nod to in our conversation, as well as headlining it as the main star in his famed "black cod with miso" dish. The fact that a little bit goes a long way in boosting your satiety index while helping to reduce your waistline doesn't hurt either.

If you're gluten-free, make sure that the miso you're buying hasn't been made with any grains that contain gluten. It's also a good idea to buy miso that's been made with certified GMO-free organic ingredients. There are many different and great local producers of artisanal miso today.

Another big way to get more umami is from natto. Made from soybeans that have been steamed, fermented, and mashed, natto is packed with probiotics and nutrients. I discovered natto almost 20 years ago while living in Kyoto, Japan. Since then I have become quite addicted to its unique taste and really love its surprising umami kick. Be forewarned: Natto delivers a strong flavor to be sure, but it's so worth it for its umami potency.

If you don't take to natto right away, keep trying. Some people find that they

R℞ *DNA Restart Health Tip #27*
Here are four ways to eat more umami from the earth.

1. Eat more tomatoes—raw, cooked, or sun-dried.

2. Use more dried mushrooms, such as shiitake and porcini.

3. Add miso to salad dressings and soups.

4. Use more tamari in your cooking and salad dressings.

need some time to adjust to the taste of natto, but if you like strong flavors such as blue cheese, you'll probably grow to like it, because it's definitely in the same family of tastes. The easiest place to find natto is at a Japanese or Asian grocery store. If there aren't any nearby where you live, you can order it from one of the many online vendors who can ship it directly to you.

Tamari—a type of soy sauce that was traditionally made from the liquid left over from miso production—is essentially liquid umami and one of the best ways to increase the depth of flavor of many a dish. Unlike other types of soy sauces, it's usually made without the addition of any wheat, so it's naturally gluten-free. Go ahead and add tamari to your next DNA Restart shopping list. For recipe ideas using tamari, see pages 232 and 240.

The bottom line: Use these traditional, umami-rich foods often and liberally to get satisfied and skinny.

EAT MORE UMAMI-PACKED DRIED MUSHROOMS

One of the most powerful and easiest ways to increase the amount of umami in a dish is to add some dried mushrooms to it. Unlike other types of food that have some umami, there are types of dried mushrooms that have a combination umami punch due to the unusual synergistic presence of both glutamate and the 5'ribonucleotide guanylate. I'll be speaking more about how to combine different foods to get the same effect in Chapter 27, Synergistic Umami, starting on page 160.

It's actually the aging and drying process (Jiro's sushi, anyone?) that really makes the umami punch power readily available in mushrooms such as shiitake, porcini, chanterelles, and morels. Vegan cuisine has been employing umami for thousands of years by using dried mushrooms such as shiitake that are rich in guanylate. Chef Corey Lee, speaking to me from his Michelin three-star restaurant Benu, extolled the virtues of dried mushrooms within traditional Asian cuisine, citing Asian vegan food as a prime example of a sophisticated cuisine that uses umami to create a layered complexity of flavor in its dishes.

Of all mushrooms, truffles possess a particularly interesting triple umami combination because they are rich in the amino acid glutamate as well as the 5'ribonucleotides guanylate and inosinate. Triple umami action in a single food is not only rare but also exceptionally powerful when it comes to the taste it imparts to every dish and is one of the reasons why truffles have been so treasured throughout most of recorded human culinary history.

Most people who appreciate truffles have usually experienced them as an infusion in a truffle oil product. Even though I swoon at the umami potential of using truffles, most of the truffle oil products available today are, sadly, completely artificially flavored. If you happen to have a small bottle of truffle oil in your pantry at home, take a minute to go and grab the bottle.

If the phrase "natural flavors" or "artificial flavors" appears in the list of ingredients, you need to toss it as part of the DNA Restart. Now that's unfortunate, because I know it was probably expensive, but the goal of the DNA Restart is to optimize your health. Why bother with any added chemical flavorings when nature has already suffused her bounty with so much delicious umami goodness right out of the earth itself?

CHAPTER 25

The Umami Potential of Aged Hard Cheese and Meat

This is the umami category that DNA Restarters will indulge in the least, as a good many of you are likely dairy intolerant and all of you will be restricted to two servings of red meat per week (reread the 1st Pillar if all of this sounds foreign to you). But since I believe that we all genetically evolved for moderate consumption of things like umami-rich meat and cheese, we're going to talk about it here. Still, rules are rules when it comes to your meal plan. Good?

Most of us are likely eating too many animal products today. And there's good reason for this: Animal products are delicious and packed with umami.

But finding ways of accessing umami within plant-based foods is a worthy endeavor as well. And chefs like Daniel Humm, whom I mentioned earlier in this pillar, were extremely inspirational in this regard; as Daniel reminded me, "sure it's a challenge to pack satisfaction into a 100 percent vegetarian dish, but it's a challenge worth overcoming . . . umami can help with that. Umami is huge."

You can overeat almost all types of foods that are umami-free, but ones that have the most umami help you break the cycle. The secret I discovered was to focus on increasing umami from all sources—so let's start with one of my favorite groups.

HARD AND AGED CHEESES

Cow's milk lacks umami, so besides using it sparingly in your coffee, if desired (but no cream is allowed), you are forbidden from drinking plain milk while on the DNA Restart. That would simply be a waste of caloric and umami potential!

Besides the relative lack of umami, milk from cows doesn't get merit points on the DNA Restart because most people on this planet just cannot break down lactose as adults (lactose is the sugar in plain milk that we discussed in Chapter 6). But the story changes once fermentation and aging enter the picture. Notice an umami pattern yet? Fermented milk and/or aged milk products like kefir, yogurt, and hard cheeses are not only better bets for people to consume because of the lower levels of lactose, but they are also full of umami.

A grated Parmesan cheese, like Parmigiano-Reggiano, goes a long way toward imparting a robust umami taste. This occurs because of the aging process. In the case of authentic Parmigiano-Reggiano, aging can take more than 2 years, and thanks to this amount of time, more of the amino acid glutamate is released—which of course triggers our umami taste perception. For an even more potent synergistic umami combination, have a look at Chapter 27, Synergistic Umami, beginning on page 160.

It's not just Parmesan that works when it comes to umami. There are many cheeses to choose from, including Cheddar, Grana Padano, etc. The harder the cheese, the more free glutamate it's going to contain—meaning more umami. This is what typically happens as cheese ages. Cheddar and blue cheese both become sharper and impart more umami the longer they're aged. If you find the taste of some aged Cheddars to be too strong, you can try either decreasing the amount you use in a recipe or using a variety that has been aged for less time. Either way, aged and hard cheeses are all very energy dense, so enjoy their umami goodness sparingly (consume no more than three 1.5-ounce servings per week on the DNA Restart).

WALKING UMAMI

Meat from animals such as cows, bison, and lamb is, of course, a great source of protein.

A genetic ancestor of yours might have discovered this firsthand one day while chasing down a wild animal such as a bison. Imagine that together with his other clansmen, they managed to spear and kill two bison. Now that they've all expended so much energy, running and spearing, they are totally exhausted. Everyone is so hungry after so many days chasing this particular herd, and from having to camp out in the cold open plains, that a collective decision is made. Your ancestor is going to have a little nibble of bison before bringing home the rest of the meat.

After all, they deserve it!

Once they got past the fur and the skin, they might have been surprised by what they tasted, or more to the point, what they did not taste.

There's in fact not much of a meaty taste at all. "What a letdown," they might well have thought to themselves. Now if your ancestor was an inquisitive type, he might have wondered if something was just "off" with this particular bison. So they decide to try a sample from the other bison as well. Same taste! Where's the meat taste that they've come to love while roasting their meats around the clan fire after a successful hunt, and more importantly, where is that "satisfied feeling thing" that usually happens after eating the roasted bison?

Okay, full admission. That ancestor of yours whom I was describing was actually me. When I tried raw meat for the first time in doing research for this book, I was completely surprised, at a loss to explain the apparent lack of a meaty flavor, and, yes, if you must know, disappointed.

There was almost no umami. None!

After 2 years of further intensive research since that day, I can now share the reason why. What you're actually tasting when you experience umami is coming from the amino acids—namely glutamate and the 5'ribonucleic acids such as inosinate and guanylate. These are not available for us to taste until the meat starts to break down to release its umami components. This can happen through a number of processes, such as smoking, cooking, drying, or salting. We just lack the genes to taste umami from raw meat. And anyway, besides the taste experience being different, we also lack the proper dentition, and the genetics behind it, to properly eat raw meat.

It's way too tough!

The raw meat I tried was actually mechanically ground up. So when you consider together the fact that raw meat is tough, its total lack of umami, and, of course, all of the dangerous microbes our gut is no longer accustomed to eating, it's not a good idea for us to be eating raw meat anytime soon.

So how should you be eating your meat?

Remember, we're after flavor here, so start by going for beef or pork that has been aged for longer, which kicks up its umami and makes it more delicious. In addition, bear in mind that once older animals are aged and cooked, they also tend to be better sources of umami—another reason beyond the ethical not to eat baby cows and leave things such as veal off the menu.

I want every bite to count toward umami because we're focusing on bigger taste and smaller portions of meat. We all need to eat less meat, but when we do eat it, let's make it count.

AIR-DRIED AND SALTED UMAMI

Given the impressive amounts of nitrates and high caloric density in processed meats, you're not allowed to have any during the DNA Restart. One notable exception: If you are incredibly DIY and happen to make your own umami-rich sausages prepared with organic, pasture-raised meat, then and only then can you enjoy a modest 2- to 3-ounce slice of salami as one of your two red meat indulgences per week. The same rule applies here—small amounts of umami-filled homemade sausage can go a long way to add taste and get you satiated.

When it comes to cold cuts such as lunchmeats—and almost everything else from the deli meat counter at the supermarket—the DNA Restart demands that you unwaveringly avoid these foods. There's no reason to eat any highly processed *mystery meats* today when there are so many other great options. The goal is to eat food that resembles or still retains the appearance of the original ingredients. Sorry, chicken and fish don't have fingers.

Many of these types of nutritionally low-quality, highly processed meats are not full of natural umami. Don't let the picture on the box or bag fool you. Most of these processed foods are packed with unnecessary sweeteners, chemical flavorings, emulsifiers, and unhealthy fats. Do yourself and your family a favor and just avoid them altogether.

FLYING UMAMI

Meat and eggs from chickens, turkeys, ducks, and pheasants are naturally good sources of umami. Certain cuts of the birds such as the breast can also be lower in saturated fats. But it's actually organic pastured chicken bones that are a great and economical way to add more umami and nutrition to the soups that you are preparing. You need to be very conscious of where your chickens are coming from. Step into any industrial farm and your senses will tell you why. Chickens are big business today with very tight profit margins, which translates sadly into thousands of birds crammed together into insanely tight spaces. Beyond the obvious ethical component, unnatural overcrowding can create many health problems for chickens. Increased rates of infections necessitate ever-increasing doses of antibiotics for the poor chickens.

Most importantly, though, all animals are what they eat (because by eating them, you're inadvertently eating what they've eaten). The fact that chickens are fed a high-grain diet drastically reduces the quality of their meat and eggs that end up on our tables. When animals such as chickens and cows are pastured, their meat tends to naturally have less saturated fat and more omega-3s. So on the DNA Restart, stick with organic chickens as well.

And again, no more than two servings of chicken a week.

Eggs are an extremely versatile food, but if you have a preexisting elevation in your LDL cholesterol levels, you can adjust the ratios of whites to yolk with an emphasis on egg whites; just remember to get enough choline. As a general guideline, regardless of your DNA Carb Consumption Category, you can eat from seven to eight eggs a week. If you're fortunate enough to have access to a good supply of pastured eggs, you should celebrate the joyous consumption of all of those choline-rich yolks as well (of course you remember the importance of choline for men from the 2nd Pillar). In fact, the yolk is where almost all of the umami lives. My concern is with eggs from chickens that derive all of their diets from grain—organic or not. Again, you eat what your chicken ate, and when it comes to eggs, if chickens are not allowed to forage for themselves, their eggs will be subpar in energizing your DNA.

CHAPTER 26

The Umami Potential of Fish, Shellfish, and Seaweed

Food that comes from the sea can be a very rich and powerful source of umami. In fact, throughout history many highly successful civilizations enjoyed cuisines that were heavily infused with umami from the sea. One of the things that can make sushi maki and other sushi rolls really delicious is a type of seaweed called nori that has often been dried into sheets that resemble black construction paper. In addition to its use in sushi rolls, nori can be easily shredded and added to rice and other dishes, thereby adding a boost of umami as well as a healthy shot of iodine.

UMAMI FISH: TOP CHEFS' SECRET FLAVOR ENHANCERS

Do you remember when I underlined the central role umami played in keeping conquering Roman soldiers energized and satisfied? (For a refresher, turn back to "Umami and the Roman Empire" on page 136 in Chapter 21.) Well, a modern version of garum can be found in the form of Worcestershire sauce—essentially a form of liquid umami that is commonly a "secret" ingredient used by many chefs to enhance the flavor of their dishes. Daniel Humm, the celebrated chef whom I've spoken about previously, shared that his kitchen thrives on the deliberate use of this fermented fish sauce in trace amounts to kick up the umami quotient in some of his award-winning dishes.

Another common addition to sauces and soups the world over is anchovy paste, which is often sold in containers resembling toothpaste tubes. Anchovy paste is particularly high in free 5'ribonucleotides, which, as you'll soon see, combines with glutamate to give an even more powerful umami taste. I'd recommend starting with adding about a pea-size amount of anchovy paste (since it can be packed with a lot of sodium) to a tomato sauce or marinade. This combination of 5'ribonucleotide inositol and free glutamate in anchovy paste increases the umami level considerably, and that's why very little anchovy paste goes a really long way toward providing umami. The most surprising thing is that when you've added anchovy paste in the right doses, you won't be able to discern a "fishy" taste in the dish you've created. If you detect "fishiness," that's a sure sign you've added too much.

If the idea of adding even minuscule amounts of fish-based sauces or condiments to your cooking doesn't sound tenable, but you're still looking to get umami directly from fish (because you should!), then cooking whole fish or buying them smoked is the best way to go.

On the DNA Restart, you should be having fatty and umami-rich fish at least two to three times a week. Fish that are rich in umami include mackerel, sardines, rainbow trout, and wild salmon. When it's prepared in the right way, eating fish rich in omega-3s has been shown to be a great way to prevent certain types of cancer and Type 2 diabetes, boost memory, and even improve sleep in children.

The positive health effects of consuming more umami-rich fish as well as nuts can begin to be evident early in life. A recent study published in the *Journal of Nutrition* found that children who consume more umami-rich fish and tree nuts that are packed with polyunsaturated fatty acids may be thinner overall as well as have less intraabdominal, or belly, fat than children who don't include these foods in their diets. And animal research data indicates that the benefits of fatty acids found in fish begin even earlier than childhood—during fetal life—and a lack thereof can seriously impair the normal development of the brain.

If you like salmon, I'd highly suggest you eat wild rather than farmed salmon. Alternating between sockeye or king salmon and smaller fish can reduce your exposure to environmental contaminants such as polychlorinated biphenyls (PCBs).

Not all fish are created equal, especially when it comes to farmed fish. Depend-

R℞ *DNA Restart Health Tip #28*
Try these five ways to eat more umami from the sea.

1. Eat more seaweed such as nori and wakame.

2. Eat more fatty fish like wild mackerel, salmon, and sardines.

3. Eat more shellfish such as shrimp and scallops.

4. Use more Worcestershire sauce.

5. Use more fish sauce like the Thai nam pla (but keep in mind the high sodium content in some brands of fish sauce).

ing upon what fish eat, they can end up having very low levels of omega-3s and natural antioxidants when compared to other fish. This is one of many reasons that farmed tilapia is not approved for the DNA Restart. Yes, it's cheap, but so is cardboard, and I don't think you should be eating that either. Besides all of the concerns associated with environmental degradation, why eat fish that are being raised on an unnatural grain-based diet and treated with plenty of hormones and antibiotics?

So no farmed tilapia while on the DNA Restart.

Conversely, fish like Atlantic mackerel and sardines are a much healthier and umami-packed seafood choice abundant in essential fatty acids such as DHA and EPA. Plus, in their canned form, they are both wild and super-reasonable economically. There's a good reason why sardines have long been a prominent staple of healthy diets: They have both glutamate and 5'ribonucleic acid inosinate, making them both healthy and delicious.

UMAMI ON THE HALF SHELL

Oysters are a great source of the essential mineral zinc and a proven way to boost your immune system. They also contain a delightful surge of umami flavor. Blue mussels, clams, scallops, and sea urchins are also particularly high on the umami hit list. Shellfish are not safe for everyone, though. As I explained in the 1st Pillar, some of you may have inherited genes that result in the uptake of too much iron from your diet. This can make you susceptible to infections by microbes found in some shellfish called *Vibrio vulnificus*. So if you're enjoying umami-rich shellfish, source them exclusively from a preferably local, small-scale reputable fishmonger who subscribes to the highest standards of hygiene and sanitation.

Also, I cannot overstate the importance of umami diversity: Do not just fall in love with one source of umami and replicate this for 28 days straight. Switch it up! The inclusion of diverse umami ingredients should be the underlying principle driving your meal plan selection on the DNA Restart.

CHAPTER 27

Synergistic Umami

For thousands of years, the best cooks have been creating the most delicious dishes using the principles behind the synergism that occurs when you mix free amino acids like glutamate with 5'ribonucleic acids such as inosinate and guanylate.

When I spoke with David Kinch, the chef of the Michelin three-starred restaurant Manresa who was also notably rated by GQ as Chef of the Year, he had the following to say: "I try to apply the principles of umami when I can. For me personally, I don't use umami for only deliciousness per se, but because in using it, I have less of a reliance on fats, which is part of my responsibility as a chef, to make food not only delicious but healthy."

Kinch himself does what he calls "simple things" in the kitchen to kick up the umami synergy in a dish. For example, he'll add small pieces of Parmesan rind or kombu to chicken soup stock.

This is what makes synergistic umami really special: Each cook has the ability to combine unique food ingredients that can dial it up or down. Including foods that have free 5'ribonucleotides can increase the umami factor of glutamate by a factor of almost 10.

It's now time that we put together everything I've told you about umami. I'll be imparting you with a list of specific foods you can combine in various ways on the DNA Restart to increase the amount of umami you experience in a dish. This will help you to feel satisfied and satiated faster, but another unexpected benefit to synergistic umami is that you'll need to use less salt in your food preparation—an important feature for people looking to limit their sodium intake.

Umami also helps us understand the reason why condiments the world over seem to be rich in this unique taste. Tomato ketchup, sriracha, barbecue sauce, oyster sauce, and nam pla are all rich in umami. In almost every cuisine worldwide, umami ingredients are used to make food delicious. Unfortunately, many of the modern incarnations of such condiments as ketchup and barbecue sauce are unnecessarily high in sugar and sodium, so on the DNA Restart you're not allowed to use either of the aforementioned. Stick to high-quality umami fish-based ingredients like anchovy paste.

Combining Umami Foods for Weight Loss:
The Synergism of Umami Taste

Good Food Sources of Glutamate	Good Food Sources of Inosinate and/or Guanylate
Aged Cheddar	Anchovy paste
Anchovy paste	Atlantic mackerel
Atlantic mackerel	Chicken
Blue cheese	Crab
Chicken	Dried shiitake mushrooms
Dried porcini mushrooms	Lobster
Dried shiitake mushrooms	Nori
Green peas	Pork
Miso	Salmon
Nori	Sardines
Parmigiano-Reggiano	Scallops
Potatoes	Squid
Salmon	Sun-dried tomatoes
Sardines	Tomatoes
Sun-dried tomatoes	
Tamari sauce	
Tomatoes	
Tempeh	
Walnuts	

For you to start thinking about umami at every meal, I've included a table that lists many common umami food ingredients. The column on the left lists foods that are good sources of glutamate, while those on the right are sources of 5'ribonucleic acids. You'll notice that there's some overlap, as some foods are good sources of both. Beyond some of the recipes I've provided for you in the book, the goal is for you to start combining foods from each column when you're thinking of preparing a meal. Using this table as a basic guide will also help you reach a new explosive level of synergistic umami combinations that will go far in keeping you full and helping you lose the weight.

So let's get started eating more umami!

THE DNA RESTART TAKEAWAYS

3rd Pillar: Eat Umami

1. Remember that umami makes you feel satiated faster and keeps you full for longer.

2. Include at least one umami-rich food at every meal.

3. Do the Umami Taste Test Experience.
   See Chapter 22 for details.

4. Stop deseeding your tomatoes.

5. Eat more miso, natto, and tamari.

6. Make sure to have umami-rich protein for breakfast to help you lose weight.

7. Eat more dried mushrooms such as porcini.

8. Add miso to soups and marinades.

9. Have umami-rich fish at least two to three times a week.

10. Use umami synergism by combining different umami foods for weight loss.

PART IV

THE DNA RESTART
4th Pillar

Drink Oolong Tea

Before he went on the DNA Restart, Chad's drink of choice was coffee, as was that of most of the people he worked with, who also happened to be fellow programmers. His appetite for mocha was in the realm of at least four cups a day, with a requisite double cream and two sugars. When a deadline was approaching, Chad often found himself with a severe bout of gastritis and a prescription for omeprazole, a medication from the class of protein pump inhibitors that help reduce the amount of acid the stomach secretes.

After becoming a DNA Restarter, Chad noticed a lot of changes. His first and immediate surprise was that replacing most of his coffee intake with oolong tea cured his gastritis, to the point that he was able to stop taking his medication completely. But the change that he was happiest about was that his abdominal girth was shrinking perceptibly. Some people at work started commenting to him about it and asking him what his secret was. They wondered whether his slimming down had anything to do with all of the tea that he was drinking. What Chad and his java-loving colleagues had never stopped to think about was that along with his four cups of coffee (and the accompanying double cream and two sugars), he was adding a whopping 850 extra liquid calories to his diet every day! What Chad alone understood after doing the DNA Restart was that oolong tea not only cut the excess calories from his life, but it also specifically seemed to target his belly fat.

All that coffee Chad was drinking combined with his sedentary lifestyle helped to fuel his weight gain and trigger his gastritis over the years. Chad now happily consumes two to three cups of calorie-free oolong tea every day, is still gastritis-free even when a deadline is coming up, and swears that his focus at work has never been better.

Yes, oolong is one special type of tea.

I've specifically developed the 4th Pillar of the DNA Restart, Drink Oolong Tea, for three important and beneficial reasons.

The first reason is that drinking certain types of oolong tea has the power to help you lose weight by stopping you from absorbing fat from your diet. I also believe that oolong tea has the power to get your DNA working for you as it specifically targets visceral, or the proverbial "belly-fat," weight. Oolong, like some other varieties of teas, has phytochemical compounds called *catechins*, which I'm going to be telling you more about in this pillar, that have the power to inhibit enzymes such as gastric and pancreatic lipases that help move fat from your gut into your body. And less body fat is something we all want.

The second important reason I'm requiring you to start drinking oolong every day on the DNA Restart is that I believe it has the power to help shift your microbiome in a beneficial direction—by promoting microbes that favor health over obesity. As I explained in the 1st Pillar, we now know that the microbiome plays a crucial role in determining your health status.

As you might remember, abnormal changes in your microbiome can lead to increased inflammation and instances of "leaky gut," which can damage your DNA. Remember the relationship between inflammation and DNA damage that we discussed in the 2nd Pillar? By lowering inflammation, oolong can help with that. It's also thought that teas like oolong have the ability to shift the microbiome toward being populated by the more beneficial types of microbes, which will certainly positively impact your health.

However, the third and most important aspect of this pillar is that you can potentially get all of these important benefits, including the ability to reduce oxidative stress, without adding any additional calories during your 28-day DNA Restart. There's no other food or beverage on Earth that's known today that can deliver that to you. Did I mention that oolong can be delicious and is a pleasure to drink? More on that soon.

So how do I know all of this, you may inquire? From traveling extensively to the top tea-growing and -producing countries in the world, as well as spending quality time in the laboratories of tea researchers whose scientific lifework centers on examining the health benefits of tea. As you're about to see, scientists have recently discovered that *theasinensins* and other compounds found in oolong tea have very powerful health-promoting actions. For the most part, people outside of Asia are usually not familiar with the oolong variety of tea. Well, that's about to totally change for you.

CHAPTER 28

The Story of Tea

No one knows for sure when people first discovered and started drinking tea from the leaves and stems of the tea plant, *Camellia sinensis*. Legend has it that one of China's first emperors, Shennong, discovered the magic of tea more than 4,000 years ago completely by accident. Now Shennong wasn't just any old kind of monarch. It's said that he was also a master herbalist, and he's still revered today by millions as the father of Chinese Medicine.

The story of the discovery of tea is that Shennong was tired after one particularly long morning of collecting medicinal herbs, so he decided to stop under a tree to rest. After setting up camp and placing a pot of water to boil over a fire, he decided that he really needed a brief respite. When he awoke from his short nap, to his complete surprise, the pot of water that had been left boiling had turned a golden yellow. Being a curious fellow, he peered inside the pot to investigate why the color had changed and saw that a few leaves must have fallen in, blown by the wind. Wondering what it might taste like, he took a small sip. Then another. And then another. No historical record can corroborate this, but I do imagine that Shennong then proclaimed this infused beverage to be simply delicious.

Tea was born.

A BEVERAGE REVOLUTION

What Shennong's nap set off was a beverage revolution that now completely encircles the globe. Today, more people drink tea than any other beverage in the world. And an incredible 7 billion pounds of tea is consumed every year. However, tea is anything but one single type of drink. Because of the myriad ways there are to grow and process the tea plant, what ends up in your cup doesn't always have the same health benefits.

For simplicity's sake, I'll be referring to tea as anything belonging to three main categories, or types, of tea that come from the tea plant, *Camellia sinensis*, and they are black, oolong, and green tea. There are also teas, like Pu-erh, that are fully oxidized, fermented, and even aged like a good wine for more than 20 years.

All specially produced tea comes with its own unique *terroir*, just like wine. This means the particular environmental factors, like soil type, irrigation, and elevation, all play a formative role in creating a very unique tea.

The biggest difference that you need to know among the main types of tea is their level of oxidation or enzymatic browning. Similar to what happens to a banana on its way to becoming overripe, the more oxidized a tea is, the blacker it may become. In the case of Pu-erh, it's not only an oxidized tea, but it actually goes through a process of fermentation as well. White tea, conversely, is often harvested with the least amount of processing; it can have the highest amount of caffeine per cup because it's made from the young tips or buds of the tea plant.

Some types of oolong tea undergo the process of roasting. This can often help take the edge off the sharp, tannin-like flavor of some teas, imparting a deep, appealing, long-lasting flavor to the roasted tea. Roasting oolongs also allows them to be aged—which can require quite a long time (some prized oolongs are aged for 25 or even up to 50 years).

Of the 7 billion pounds of tea consumed every year, most is the black or fully oxidized variety. This isn't the type of tea I'm prescribing for your DNA Restart, though. There are very important reasons that you'll see for my choosing oolong above all other types of tea for you to drink every day. Chief among these is a recent scientific research discovery involving a very special group of polyphenol compounds that are found in oolong tea, called *theasinensins*. Theasinensins are a big deal, and you'll be increasing your consumption of them by brewing oolong the DNA Restart way every day.

Before I get into more of the important scientific details concerning the 4th Pillar, Drink Oolong Tea, for your DNA Restart, I want to bring you up to speed about the different types of tea, and oolong in particular. When it comes to tea, there's a lot to take in, as there are hundreds of different varieties. On the following pages, I've provided you with what you need to know to effortlessly navigate the formidable world of tea, particularly oolong, so that you can immediately join the millions of people, predominantly in Asia, who drink and reap the health benefits of oolong tea every single day.

Within each of the three main categories of tea—black, oolong, and green—it is estimated that there are hundreds of different subtypes. The fact that most people drink and, therefore, are most familiar with the fully oxidized type, black tea, is really more an accident of history than anything else.

Think of this oxidizing process as a form of preservation. As the tea becomes fully oxidized, it turns black because of a chemical process that tea processors employ after harvesting. Historically, this ensured that when tea was shipped to European ports from distant lands, such as China, it would arrive preserved thanks to the oxidization process. Remember, before jet travel changed everything, journeys spanning that geographic distance could take upward of months.

Yet even fully oxidizing tea did not always guarantee its survival during the voyage. If crates or bundles of tea became wet or waterlogged, they would often spoil. That's one of the reasons why we ended up today with a tradition of flavored black teas, such as the ubiquitous Earl Grey. The oil of bergamot that was added to make the first Earl Grey was initially added to cover up the fact that the tea that was being sold was of an inferior quality, which is polite speak for a little rotten.

BLACK TEA

Although many people around the world have grown happily accustomed to the taste of black tea, from a health benefit perspective, one of the biggest drawbacks of its oxidation process is that it creates a brew that's much more astringent and bitter. (For comparison, green tea is not allowed to oxidize at all, so it's at 0 percent oxidation; an English-style black tea is allowed to fully oxidize, which is why it's at 100 percent oxidation.) To soften the potentially robust-tasting qualities that come from a tea that has been 100 percent oxidized, people who enjoy drinking black tea the world over have adopted the custom of either adding dairy in the form of milk or cream and/or adding a sweetener like sugar.

As you might remember from the 1st Pillar, most people in Asia cannot consume milk as adults. This is probably another reason that the English style of serving black tea with milk is not as popular in China. People there aren't able to temper its assertive taste with milk or cream. But when I spoke to many practitioners of Chinese Medicine in Asia, they told me that they did not advise their patients to consume the English style of black tea every day because they believed that it could be too "warming" or "harsh" on the body for some people.

One of the other drawbacks of adding milk to black tea is that it not only neutralizes the bitterness but also strips away some of the health benefits that black tea could provide. Some of the proteins that are naturally found in cow's milk, such as α-lactalbumin and ß-lactalbumin, interact and bind with the health-promoting compounds in the black tea, making them less available for your body to use.

GREEN TEA

As I searched for the type of tea that would have the maximum health benefits, my thoughts naturally turned to green tea varieties, since there's a lot of scientific research testifying to their various health-promoting qualities. There was just one slight problem. The more I personally experimented with green tea over the years, as well as consulted with tea researchers, there was one recurring issue that was hard to get around. And that's the fact that consuming many cups of green tea a day is okay for some people—but not all, since it can cause nausea and gastrointestinal upset from the natural phytochemicals in the plant.

My own experience with green tea began when I lived in Kyoto, Japan, for a year after completing my bachelor of science degree, and before starting graduate and medical school. I've enjoyed drinking green tea ever since, but I do find that I can't tolerate it at full strength every day. Just to be clear, we're not talking about the low-quality variety that's often packaged in tea bags and passed off as green tea in the United States and abroad. And it has to be obvious that I'm not talking about Snapple Green Tea here. Rather, I'm referring to a loose-leaf, full-strength brew that has the power to protect your DNA.

As you remember from the 1st Pillar, the last thing you want to do is to inflame your gut, especially now that you've removed offending agents like those nasty emulsifiers from your diet. Interestingly, I heard from many practitioners of Chinese Medicine that they likewise believed that drinking too much green tea can be irritating for some people. When I asked for further clarification, I was told that they thought green tea could be too irritating for certain people depending upon their constitution.

They could tell I was hungering for more details, so they graciously referred me to the most complete and comprehensive medical book ever written in the history of Traditional Chinese Medicine. Recognized by the United Nations Educational, Scientific, and Cultural Organization (UNESCO) for its significance, the text I was referred to was the 16th-century medical text called the *Ben Cao Gang Mu* (*Compendium of Materia Medica*) (本草綱目). The author, Li Shi-Zhen, spent 27 years and used more than 900 previous texts as sources, including his own experience as a physician and herbalist, to create this text.

Amazingly, Li Shi-Zhen's description written almost 500 years ago echoed and matched what I've heard from so many of my own patients:

> "Long-term consumption of [green] tea by individuals with weak body and blood will cause severe irritation of the stomach . . . "

Even though green tea is consumed often in Japan and can even be acquired, hot or cold, from vending machines on nearly every street corner, many Japanese have reported similar gastrointestinal issues to their own physicians as well. This is even recognized by the Japanese tea researchers, who study green tea and publish information on its health benefits. They actually don't personally drink it daily either, as they've come to recognize that drinking many cups of green tea a day can be rather hard on the stomach, especially if it's consumed at full strength.

One helpful loophole that's been identified for green tea has been to consume it as green tea extract in pill form. As you might remember from the 2nd Pillar, though, I'm very careful about recommending any supplements and vitamins. That's because it is difficult to guarantee what's actually inside your nutraceutical capsule. So when it comes to tea, why take a pill, since as you're about to discover, there's a far superior alternative?

When I looked inside the cups of tea researchers, I was quite surprised not to find any of them drinking green tea either, but something altogether different.

They were drinking oolong.

CHAPTER 29

Why Drink Oolong?

As I traveled around the world to research the health benefits of tea from Sri Lanka to Japan, I discovered that there was a buzz of excitement around a unique family of polyphenol compounds that are found in oolong tea called *theasinensins*. When I asked tea researchers why they were drinking oolong and not green tea, they said that besides the fact that it is easier on the stomach, oolong tea is the richest tea in polyphenol compounds, many of which are absorbed into the body while you're drinking. And the polyphenol compounds from oolong also help to shift the microbiome in a positive direction. These unique compounds are especially enriched in oolong because of the distinctive way this tea is produced. Unlike other teas such as white, green, or black, oolong is only partially oxidized. This partial oxidization process does an exceptional job of capturing, creating, and maintaining the unique polyphenol compounds.

Research has consistently shown that new habits are hard to form, but once you get through the first 28 days of your DNA Restart, you're much more likely to make them stick. It also takes time for your DNA to adjust to your habits. Drinking oolong every day will help in the process of turning "on" genes that promote health and longevity, and at the same time turning "off" overactive genes that are detrimental. The dialing up or down of the volume of certain genes ultimately determines the outcome between the competing forces of disease or health and longevity. That's one of the reasons for the length of your DNA Restart—it can take time to reprogram your DNA. And sipping oolong every step of the way will help.

Now before I outline how you're going to be using oolong during the next

I'm usually only a coffee drinker and never really liked the taste of tea. Since it was part of the DNA Restart, I gave oolong tea a shot, and I really felt so much better after I started drinking oolong tea every day. I could focus better at work, and I never felt jittery, even if I drank it in the evening—something that I just can't do with coffee.

—Andy, 52

28 days of your DNA Restart, I want to bring you into the very exciting and relatively unknown world of tea research. Once you start drinking oolong, you'll begin to feel the difference, not just as a tool for weight loss but for the myriad other systemic benefits it provides as well.

By drinking oolong tea every day, you will be joining millions of people who for thousands of years have recognized its health-promoting benefits. Cutting-edge science is only now beginning to understand and unravel that oolong is good for your DNA.

OOLONG HAS MORE POLYPHENOLS THAN ANY OTHER TEA

Since so much of the exciting research work with oolong has only just begun, you likely haven't heard that much about it yet. Many of the new polyphenol compounds discovered so far are found only in oolong, so I decided to visit some of the world's leading tea researchers who are doing the most cutting-edge research so that I could learn even more about the health benefits of this ancient brew. To properly do that, I knew I needed to go to the one place that has some of the most unique and special oolong producers as well as researchers in the world.

So I boarded the first flight I could find to Taiwan—one of the world's top producers of high-quality oolong teas. After landing, I headed to Taichung on a super-sleek high-speed train hurtling down the tracks at 186 mph.

I traveled first to Taichung because it's the home of the Tea Research and Extension Station. Dating back to 1903, the research center is surrounded by beautiful green rows of tea plants. Initially, it was established as a place to help farmers and producers improve the quality of tea being produced in Taiwan. Since their founding more than 100 years ago, the research institute has expanded their mandate to include advanced scientific work exploring the health benefits of tea, especially of oolong. They currently are conducting more than a dozen research projects that are focused almost exclusively on studying the unique family of polyphenol compounds called theasinensins.

Most of the natural phytochemicals that can sometimes cause nausea and gastrointestinal upset are actually found in some kinds of green tea and much less so in oolong. It shouldn't surprise you that green tea could cause nausea and irritation for some people, because I'm sure you remember from the 2nd Pillar that plants often produce phytochemicals as poisons to stop you or anything else from eating them.

Keeping tea "green" is actually the easiest method of tea production. When it's first harvested and still green, as in the Japanese style of sencha, processors apply heat to the leaves to stop any oxidation. Heat works because it inactivates an enzyme called *polyphenol oxidase*; when this enzyme is inactivated, the leaves maintain their deep green color. Many years ago this process was done by briefly cooking tea leaves in a very big wok similar to what is customarily used in Asian cooking. Today it's often done within giant machines that look like commercial dryers or laundry equipment.

As I mentioned earlier, the biggest difference among the types of tea is the level of oxidation. When tea is first harvested, the fresh leaves contain phytochemicals such as polyphenols, which include *flavanols, flavadiols,* and *flavonoids.* In addition, there are a host of other compounds called *catechins* that are plentiful in green tea and include (–)-epigallocatechin-3-gallate (EGCG); (–)-epigallocatechin (EGC); (–)-epicatechin-3-gallate (ECG); and (–)-epicatechin (EC). Tea leaves also naturally contain antioxidant polyphenols such as kaempferol, myricetin, and quercetin.

The big difference in which types of compounds you find in the different types of tea—and the levels at which they appear—is dependent upon the degree of oxidation; since green tea is the least oxidized, it correspondingly contains the most EGCG, EGC, ECG, and EC. This is also why it's the most irritating.

The drawback of inactivating enzymes such as polyphenol oxidase so early in the tea production process, as done when producing green tea, is that plenty of the compounds that can cause an upset stomach for people still remain. So what happens then if you don't apply heat to green tea leaves? Well, the enzyme polyphenol oxidase starts to oxidize and convert catechins to *quinone,* and then to other compounds such as *theaflavins* and *thearubigins.*

Eventually, if you let this process continue almost indefinitely, you get what is known as an English-style black tea. As you'll recall, black tea is one of the most oxidized types of tea, and this high level of oxidation causes many of these new molecular compounds to eventually start to unwind and unspool. The chemical process is called *oligomerization,* and these molecules are then polymerized to form the theaflavins and thearubigins. That's why black tea consequently has the highest amount of this oxidized tangled molecular mess.

When this happens, the molecules join up together in a big molecular jumble. Imagine a large pile of unspooled and tangled wool from threads of different colors and thicknesses, and you'll get the picture. It's these new large piles of tangled compounds that many people find too harsh to drink straight up, prompting them to add milk and sugar.

Oolong is considered by most tea manufacturers to be the most challenging type of tea to produce. It requires a really skilled tea artisan or master to get just right. The simple reason for this is that unlike green tea that is merely heated, or black tea that's allowed to completely oxidize, oolong is a finicky compromise between the two and requires many more careful, attentive steps along the way.

What happens when you stop the tea oxidation process somewhere in the middle? You'll end up with a very happy molecular medium—not too much EGCG, EGC, ECG, and EC or too many theaflavins and thearubigins. In fact, the reason why most oolongs have the highest amount of polyphenol compounds—more than any green and black varieties of tea—is because the oxidation process is stopped at anywhere from 20 to 80 percent oxidation. Allowing the natural enzymes in the tea leaf to keep changing the polyphenols after harvesting is another reason for the presence of these unique compounds. Further, unlike in a black or fully oxidized tea, more of these polyphenols are easier on the body because they haven't become all tangled together like they do in the black tea production process.

There is also an artisanal, skillful balance required in the art of producing oolong. It can take anywhere from 36 to 48 consecutive hours of an impressively labor-intensive process to produce a good oolong. The reverence for the immaculate implementation of this time-honored tradition is exemplified in the Taiwanese saying "What is done in the processing of the fresh leaf is going to show up in the cup."

It is this very distinguished processing that gives oolong tea its unique chemical polyphenols and properties that are unlike every other tea. It's during the processing that new polyphenol compounds constantly emerge. Some of these naturally produced compounds have incredible potential health benefits for you.

While in Taiwan I also visited the last remaining traditional small-scale oolong roaster on the entire island. This revered master roaster produces roasted oolong by a method that hasn't changed in thousands of years. To roast tea in the old and traditional manner, wicker baskets are filled with incredibly fragrant tea and are roasted by hand over slowly burning bamboo charcoal embers. Sounds romantic, right? But roasting oolong the traditional way is back-breaking and unforgiving work. Get the

R℞ DNA *Restart Health Tip* #29
Here are the top seven reasons you need to drink oolong tea!

1. Blocks dietary fat absorption
2. Targets belly fat
3. Speeds up your metabolism
4. Reduces inflammation

5. Protects your DNA
6. Boosts your creativity
7. Reduces anxiety

temperature wrong, and it's very easy to turn a sweet and fragrant oolong to something akin to burnt toast. To make sure that doesn't happen, this one and only master roaster works with his apprentice, who takes over after 18 hours straight of roasting.

In the hot, darkened room in Taiwan where I had the pleasure of observing the roasting firsthand, I noticed that the woven baskets that were used to roast the oolong were lined with a white crystalline powder. I asked what this powder was, and the roasting master encouraged me to taste it. So I did just that. It tasted mind-numbingly bitter.

The taste transported me back almost instantaneously to years earlier in my organic chemistry laboratory days when I was tasked with the job of extracting caffeine from tea leaves. It turns out that the roasting process of some oolongs is a natural way to reduce the amount of caffeine present. If, because of the DNA you've inherited, you're sensitive to the stimulatory effects of caffeine, then on the DNA Restart you should be consuming only roasted oolong, because it naturally has less caffeine.

CHAPTER 30

Oolong, Polyphenols, and Oxidative Stress

Research at the Taiwanese Tea Research and Extension Station has found that when tea plants are stressed either by high altitude, which exposes them to higher amounts of UV radiation, or drought and even pest infestation, they produce more polyphenols.

These findings reminded me of my research trip to the Altiplano in Peru, which I delved into in the 2nd Pillar, where potatoes grown at high altitude responded to having more UV radiation by increasing their amounts of phytochemicals; these same phytochemicals, when consumed by humans, are believed to mop up oxidative stress. It's also the reason that some oolong teas that are made from summer-grown tea leaves usually taste much more bitter. In places like Taiwan, that's the case because of the sweltering hot and humid summers. Additionally, the tea plants have to cope with many more insects and fungi at this time of year, which increases further the amount of polyphenols they produce.

Let's look in a little more detail at some of the health benefits from drinking oolong that you'll be enjoying over the next 28 days of your DNA Restart.

OOLONG AND YOUR WEIGHT

The reason that you'll be drinking oolong every day of the DNA Restart is that the incomparable health benefits that oolong imparts will not add one additional calorie to your diet. There's no other drink on the planet that can offer you that.

For example, the myriad unique polyphenols in oolong as well as the caffeine will help to dial up your basal metabolism. This can have the effect of raising your body's core temperature, which means your body will be burning off more energy from your food. In addition, caffeine from oolong can help stimulate your body to burn fat by turning on your sympathetic nervous system. But unlike coffee that also contains caffeine, other compounds found in oolong mitigate some of the negative symptoms that can be associated with coffee intake in some people, such as

175

increased anxiety. That's one of the reasons the DNA Restart limits your coffee intake to two cups a day, and that's without any cream or artificial sweeteners.

Both the caffeine and other compounds in oolong are thought to turn on your brown fat, which, unlike other types of fat in your body, helps reduce your weight by burning sugars and fats. It was once thought that only young babies had a substantial amount of brown fat. But as I explained in my first book, *Survival of the Sickest*, even if you're an adult, you can have your brown fat activated and subsequently burn more calories if your body experiences prolonged cold. The drawback of using cold temperatures as a dietary tool (I was actually involved in research trying to do just that) is that it's of course extremely uncomfortable.

So if you're not a fan of freezing for hours on end even if you are rewarded with some extra brown fat, there is another way—and that's to drink more oolong instead!

Another way that oolong is thought to help keep you thin is that the catechins within it are believed to help stop you from manufacturing fat cells, known as white fat (the kind that most people want to have less of).

These are just a few of the reasons why drinking oolong tea long term is thought to help you: It may tweak your DNA to help you burn off more fat by raising your basal metabolic rate as well as cause your body to make less white fat, which then lowers your overall level of inflammation.

Yet not all research studies on tea have demonstrated positive results when it comes to lowering body weight and keeping the pounds off. Still, I was seeing both weight loss and maintenance in myself and my patients who were drinking oolong.

Intrigued, I've spent a lot of time with tea researchers in Taiwan and elsewhere discussing this issue, and the one recurring theme that emerged was that many of the studies didn't actually use brewed tea. What they were using was green tea extract in pill form, which I'm not a proponent of, as I've told you previously. The pill form of green tea of course lacks many of the phytochemicals that are found only in oolong, but it also even lacks some of those that are naturally found in brewed green tea.

This is why, during the next 28 days, you will only be drinking oolong tea that's brewed the DNA Restart way so that you can get the maximum amount of phytonutrients such as polyphenols out of the oolong tea leaves.

The DNA Restart is about eating food that our genes evolved to thrive on—and that includes drinking real, properly brewed oolong tea. As a DNA Restarter, you just don't do "extracts."

OOLONG AND DIETARY FATS

Another interesting thing that drinking oolong will do is help you absorb less fat from your diet. Right off the bat, that's remarkable, because the less fat you absorb, the less extraneous energy, or calories, you're taking into your body overall. By absorbing less fat, your body is also now forced to rely more on your own fat stores. This is important because your body will learn to shift your DNA from a mode of pure fat absorption into one of fat utilization and burning.

As you learned in the 1st Pillar, Eat for Your Genes, your DNA is actually primed to help you absorb all the fat you can from your diet. And that's because for most of your genetic ancestors, fat was incredibly hard to come by. Wild animals are of course by nature rather lean, just like their domesticated pastured descendants. In order to ensure our collective survival as a species, we evolved DNA that encodes for genes that help us absorb fats that may be present in our diets.

Without the functional DNA to acquire, assimilate, and properly absorb fat, your ancestors would not have survived long and you wouldn't be reading this right now. This is really what your DNA has cared about all along: your survival. It's not our DNA's fault that it hasn't quite caught up to our modern era of hypercaloric abundance. Our natural world has been turned upside down dietarily, and most of us are not getting the proper nutrition that we need. Today, most people eating a Western diet have no problem getting more than an ample supply of processed sugar and fats.

So what does oolong do to help out our fat-hoarding DNA? It blocks pancreatic and gastric lipase. These important enzymes help to break down the fats you consume into fatty acids and glycerol, making them superavailable for your body to absorb. When groups of polyphenol compounds that are found in oolong, such as theasinensins, inhibit these enzymes, they stop fats from being absorbed into the body. That's why on the DNA Restart you should drink your oolong with your more substantial meals to maximize its inherent fat-blocking properties. Drinking oolong every day then forces your body to rely on and use more of its supply of stored fat, as you absorb less fat naturally from your diet. As you're going to see soon, there are components of tea that also work as *prebiotic* compounds that will help shift your microbiome in a more healthy direction.

THEASINENSINS MEAN BUSINESS: HOW OOLONG TACKLES INFLAMMATION

As you might remember from the 2nd Pillar, Reverse Aging, the various phytonutrients found in certain foods can have an enormous impact on protecting your DNA and body from oxidative stress such as reactive oxygen species (ROS). Theasinensins are technically not a phytonutrient because they are only found in oolong after it's been partially oxidized.

This doesn't stop theasinensins from packing quite an antioxidant and anti-inflammatory punch! That's not all they do, as they've been shown to stop certain types of fats within your body from becoming oxidized as well. This is a great strategy to slow down the aging and inflammatory processes within your body in general.

In research from Japan, theasinensin-A and theasinensin-B, which are both found in oolong, were found to inhibit an enzyme called *cyclooxygenase-2*, or COX-2, that's made from your DNA and involved in the inflammatory process. This is the same enzyme that aspirin also targets; oolong essentially destroys the enzyme, lowering inflammation as a result.

Theasinensins and other compounds found in tea can also bind iron in your diet. These compounds from oolong will usually bind nonheme iron, the kind that comes from vegetable sources, and not let your body absorb them. As you might remember from the 1st Pillar, Eat for Your Genes, too much dietary iron can cause an oxidative strain on the body, so oolong's iron-binding abilities can be wonderful for people with excess iron. If, however, you're planning on doing the 28-day DNA Restart as a vegetarian or vegan, avoid drinking your oolong with your meals. This precaution should allow you to still get enough iron.

Whatever you do, though, you are positively forbidden from adding any milk or dairy substitutes to your oolong (remember that according to the 2nd Pillar, you shouldn't be having any nut milks anyway due to the frequent use of emulsifiers in their manufacturing process), as this can potentially interfere with the absorption of many of the health-promoting polyphenols. Plus, oolong tastes best straight up.

YOUR GUT MICROBIOME AND OOLONG

You can think of your microbiome as your very own family farm. Depending upon what you eat, you can have incredible variations in your gut microbiome. The differences among microbiomes can be so significant that you can even tell monozygotic, or "identical," twins apart by looking at their microbiomes. But just as the properties of the soil can affect the nutritional qualities of the plants on your farm, your gut microbiome can radically change your health as well.

For example, as you may remember from the 1st Pillar, Eat for Your Genes, eating foods with lots of emulsifiers can negatively impact your gut microbiome. It's not just emulsifiers, though, that can affect your gut microbiome. When researchers studied children from European countries who were eating diets higher in animal protein and carbohydrates, their gut microbiome resembled a monoculture crop field that didn't have that much variability in microbes. Picture row upon row of identical corn plants (instead of microbes), and I'm sure you get the picture.

When the same researchers sampled the gut microbiome of children from Burkina Faso, in Africa, who were eating a diet that was low in fat and animal protein and high in dietary fiber, they found a gut microbiome that was much more varied and had more of the strains of microbes from Bacteroidetes. What we eat has such a big impact on the composition of microbes that make up our gut microbiome that children living in Burkina Faso of a similar age produced more of a short-chain fatty acid, called butyrate, which nourishes the enterocytes, cells that line the gut.

In a way, you and your microbiome are one. Since much of what you consume is modified by your microbiome, the bad news is that even if you're not overweight, but are eating a diet that's high in fat and red meat, you're at a much higher risk for colon cancer and cardiovascular disease, and more than likely, your gut microbiome plays a very big role in that process.

The good news, though, is that so far, in the young and burgeoning scientific field of the gut microbiome, research is indicating that a change in diet can cause a corresponding change in the microbes that call your gut home.

The microbiome also seems to play a very significant role in obesity. For example, research studies have shown that the microbiome of obese and lean individuals can be drastically different. And our microbiome is likely intimately connected to body weight, as preliminary animal data is suggesting. When the microbiome of the gut from an obese mouse was transplanted to a thin mouse, that thin mouse became obese as well.

So it seems that the microbiome you cultivate within your gut can change the way in which your DNA behaves, in an epigenetic manner, by turning the volume up or down on your genes. The impact of eating more fiber, for example, can lower the risk of colon cancer by not only shielding the gut from toxins you ingest every day, but also by offering your microbiome a plethora of phytochemicals that, when broken down, can change the epigenome.

That's why drinking oolong tea every day is important, since it contains catechins. Mice that have been fed a high-fat diet with catechins (versus the ones fed a high-fat diet without supplemental catechins) show an improvement in the type of genes that are being used from their DNA in a way that leads to their fats being burned more efficiently. Some laboratory experiments have also found that catechins derived from tea can work like a prebiotic, helping to shift the gut microbiome in a more healthy direction.

So when you take your gut microbiome into consideration, you are evidently so much more than what you eat. You're also the trillions of microbes that live inside your gut.

Given that, isn't it time that you brew them some oolong tea?

CHAPTER 31

Brewing Oolong
the DNA Restart Way

The goal of the 4th Pillar, Drink Oolong Tea, is to consume between two and four cups of oolong tea per day, which will give you the maximal benefits during your 28-day DNA Restart. If you find yourself very sensitive to caffeine, I suggest that you try drinking your oolong before 3:00 p.m. every day or that you decrease the brew time.

As I wrote about earlier in this pillar, roasted oolong is another wonderful oolong option that naturally contains lower levels of caffeine. I spent a lot of time working with third-generation Taiwanese tea masters perfecting and revising the most accessible way for you to prepare oolong tea at home today. The main goal was to make sure that you get the most polyphenols and other health-promoting compounds into your cup.

The first rule of making the perfect cup of oolong is that you are never permitted to add milk or sugar. I touched on this before, but it bears repeating because of the important science behind it. Milk will bind theasinensins and some of the other precious polyphenols in oolong, making them less accessible for your body to use. This makes drinking oolong pointless.

The second rule is that you should make every effort to procure the highest-quality oolong you can find—either at a local health food store, supermarket, or even online. What you should watch out for are those low-quality tea bags that are labeled as being oolong, but are anything but. This is similar to our discussion in the 2nd Pillar, when I gave you the reasons behind why you need to use extra-virgin olive oil on the DNA Restart.

Believe me, your DNA will thank you.

A way to ensure that you're getting better-quality oolong is to look for tea bags where you can actually see the whole leaves unfurl once the bag has been submerged in water. Oolong tea leaves almost always need room to expand. That's why some of the best oolong tea bags look almost empty before they're wet, as the oolong hasn't expanded and unfurled yet.

THE BASIC BREWING METHODS

Now let's turn to the actual brewing process. We'll begin with the "hot" DNA Restart way of brewing oolong, which is the easiest way to start. To prepare oolong using the hot method, place your oolong tea bags in a teapot or heat-safe jug. Use 6 ounces of water for every tea bag you use. Now wait for your water to boil (the optimal temperature is 195°F), and after it has, wait 1 to 2 minutes for it to cool slightly. Now pour the slightly cooled hot water into the teapot or jug and let it steep for 1 to 2 minutes. Finally, remove the tea bags. You can either drink your oolong now or save some of it in the fridge to have later.

The other method that I've tested both with tea masters and many DNA Restarters is the "cold" DNA Restart way of brewing oolong. You can use the technique listed above, but instead of using hot water, simply use cold water and let your oolong brew undisturbed for at least 2 hours or overnight in the fridge. Research has now shown that hot or cold, brewing oolong the DNA Restart way will get the most health-promoting compounds into your cup. Happy brewing.

FANCY OOLONG BREWING METHODS

For the more serious-minded tea-brewing people out there, you can buy whole loose-leaf oolong, which can also be used for more than one brew. I will be providing instructions on how to make fancy oolong using loose leaf, but I'll keep things relatively straightforward. Loose-leaf oolong is almost always of higher quality than the bagged varieties, as it's also easier to visually inspect the tea leaves to ensure high quality. Highly coveted and virtually unattainable, the absolute highest grade of oolong doesn't even leave the island of Taiwan and is usually spoken for long before it's harvested. But there's still plenty of excellent-quality loose-leaf oolong available.

Most of the novice oolong brewing mistakes involve either using water that's too hot or too cold. The other issue is leaving the tea to infuse in the water for way too long or not long enough.

I have simple solutions for both of these matters that will make brewing your own loose-leaf oolong at home both fun and delicious.

If oolong is brewed with water that's too hot, it can become overextracted, which can often affect the taste. When this happens, you can lose out on all of the subtle taste within the tea, which makes it less pleasant to drink. If oolong is brewed with water that's too cold, many of the compounds in the tea leaves that are health promoting will not be able to properly seep into the brewing water.

The optimal temperature for brewing your oolong is 195°F, but don't worry if you don't want to use a thermometer, because the DNA Restart has you covered with a simple technique. The tea masters I worked with in Taiwan visually inspect the water as it gets hotter and wait for what they call shrimp eyes before they pour it onto their oolong. This is when, just before water comes to a boil, you'll see two small round bubbles close together that resemble cartoon-shaped shrimp eyes.

To make your oolong, though, you're not going to have to use a thermometer or even look for shrimp eyes, unless you really want to. What you're going to need are a kettle and two vessels that pour easily. One vessel will be used to cool down the boiling water, while the other will be used for you to brew your tea within. While you're waiting for your water to boil, prepare the second container, which can be a teapot or even a mug, by putting the tea inside the vessel in the correct amount. Do not use a small metallic tea ball for your oolong. If you'd like to use a strainer, I'd suggest a large mesh strainer that's usually the size of the inside of a mug, as this will allow the tea leaves to open fully and enrich your brew accordingly.

You should premeasure your teapot or mug to find out its capacity by using a measuring cup. As a general oolong brewing rule, use ½ to 1 teaspoon of loose tea for every 6 ounces of water. Over time, you may decide you like a stronger or weaker brew by adding more or less loose oolong leaves, but begin on the smaller end of the spectrum.

The water preparation technique works as follows: Wait for your kettle to boil your water. Many kettles stop boiling at about 200°F for their first boil run. Once it's boiled, carefully pour out the amount of water you will need (3 to 6 ounces, depending upon your brewing preferences) into the first container. When you start to feel this container getting warm, again carefully pour the hot water onto your oolong (3 to 6 ounces per serving cup), which will be loose leaf. For the *first* brew of oolong, the steeping time is 60 seconds. After 60 seconds, pour the oolong-infused water into a drinking vessel, leaving the whole tea leaves in the steeping vessel. Your oolong is now ready to be enjoyed.

By using two containers, the first one briefly, you will naturally be lowering the temperature of the water. Some hot water kettles today come with a temperature setting, so if yours does, go ahead and set it to 195°F, which will save you the transferring step that's otherwise required to cool down the water temperature.

One of the benefits of using loose-leaf or a high-quality bagged oolong is that you get many more brews out of the same leaves than with any other type of tea. If you'd like to be exact and insist on using a timer, the most accurate brewing times for extra brewing are as follows: For the *second* brew, the steeping time is 70 seconds; for the third, it's 80 seconds. You can keeping adding 10 seconds for each additional brew cycle. Remember to always pour the oolong-infused water into a drinking vessel after every infusion. For most types of high-quality oolong, after five rounds of hot water infusions, the health properties and distinctive oolong tea flavor will have been maximized.

I've gotten many questions from DNA Restarters regarding the best type of water to use for brewing your oolong. Before you think about that, though, pay attention to the water-boiling vessel itself. Make sure your kettle doesn't have a big buildup of scale, since that can really affect the taste of your oolong. From my research I cannot recommend using distilled water or water that's too hard—somewhere in the middle provides the best results. Further, regarding the pH of the water, the best results are from water that has a level between 7 and 8. It's fun to experiment yourself to see what you like best.

There are also much more advanced brewing methods, including using miniature *Yixing* clay teapots that resemble a child's tea set. Then there are *gaiwans*, which are great for single or multiple servings. These are not necessary for your 28-day DNA Restart, but once you start feeling the difference from drinking oolong, you'll quickly find yourself getting more into brewing.

Now if you simply want to drink oolong cold, wait for the oolong you've brewed to cool down enough to put it into the fridge. When I brew oolong, I usually make more than I can drink in one sitting and have the rest sometime later in the day.

THE DNA RESTART TAKEAWAYS

4th Pillar: Drink Oolong Tea

1. Limit your coffee intake to two cups a day (no cream or sweeteners allowed).

2. Remember that oolong tea is the richest polyphenol-packed drink that's naturally calorie-free.

3. Take advantage of the fact that, unlike green or black tea, oolong is easy on the stomach.

4. Drink two to four cups of oolong every day.

5. Do not ever add milk or sugar to your oolong.

6. Oolong tea protects your DNA from oxidative stress.

7. Oolong tea is good for your gut microbiome.

8. Drinking oolong tea specifically targets belly fat.

9. The polyphenols and other compounds found in oolong tea can reduce inflammation.

10. Compounds found in oolong tea block dietary fat absorption.

PART V

THE DNA RESTART
5th Pillar

Slow Living

Having recently returned to working full-time, Kelly, a 39-year-old mother of three very high-energy boys, often reminisced about the quiet and intimate candlelit dinners she'd shared with her husband in the early years of their marriage. She remembered how they used to eat slowly, savoring their food and enjoying one another's company. Far from being unhappy, though, she relished having a full and at times very loud home. It was just that with so many distractions, and having to keep up with the incessant stream of work-related e-mails, there was almost no meal that didn't involve having a fork in one hand and a phone in the other. And before she knew it, she had polished off the entire contents of her plate without even registering the first bite. Not too surprisingly, this would lead her to mindlessly taking second and sometimes even third helpings of whatever happened to be on the table. But it wasn't just Kelly who was spending mealtimes this way; her kids and husband would polish off their dinners while simultaneously texting, checking e-mail, or trying out the latest app on their phones. Plus, meals were often cut short to accommodate some scheduled after-school activity. Kelly was delighted to recount to me that what she most appreciated on the DNA Restart was, "No texting! No phone calls! No distractions! Just sitting and eating delicious food as it was meant to be eaten. I never realized the amount of time as a family we spent completely disconnected even though we'd all be in the same room."

Using the fifth and final pillar of the DNA Restart, Kelly dramatically changed her family's mealtime habits, which brought much-needed moments of enjoyment to their lives at mealtimes. And the unexpected perk? By applying the principles of the 5th Pillar, Kelly and her family learned to be consistently satisfied with smaller servings at every meal.

Slow is the final secret to recalibrating your life to maximize your DNA. Your DNA has the capacity for an immense amount of change. Think of it right now as an orchestra that's awaiting you, the conductor, to stand up, wand in hand, and lead it through the symphony of your life. Through the first four pillars of the DNA Restart, I have provided you with the "score" for this orchestra, by giving you scientifically based guidance to find out for yourself what and how to eat and drink, and how to stop and reverse genetic aging. Now it's time for you to pick up that orchestral wand for yourself. But make sure to do so slowly.

The main reason that our DNA today is out of sync with the lives we live is speed. We never evolved to be continuously sleep deprived, always eating quickly, and spending so much time away from the ones whom we deeply love and are loved by. All the stressors caused by modern living have likely already taken a toll on your DNA. We know clearly that continuous stress puts us all in an unhealthy state that will even change the DNA we pass on to the next generation.

The exciting and very challenging time I spent training and working as a physician taught me one important thing: The human body can endure only a certain amount of stress, for so long. Research has now conclusively proven that when you combine lack of sleep with poor eating and minimal exercise, you are almost guaranteed to get fat and really sick.

But why is that?

Our ancestors spent most of their day acquiring, preparing, sharing, and eating their food. Modern living has solved many of the problems our genetic ancestors faced in acquiring and preparing foods, but in exchange it's created a much bigger problem. *We have all forgotten how to eat.*

In my own life, for many years I prided myself on being able to relegate eating to just another task, to be expertly combined with walking, talking, and even running down busy hospital corridors. And because I've heard about this phenomenon so often from my patients, friends, and family, I know that I'm not alone when I tell you that I was lacking in the most basic of human skills: slow eating.

The 5th Pillar of the DNA Restart is a guide to get you back in tune with the unique genomic wisdom that we all have inside us. The speed at which we eat, work, and live is in direct conflict with all of our various genetic ancestors. We have all turned our backs on what they have to say about how to live and eat right. And the consequences are quite apparent. Look around and what do you see?

Epidemics of obesity, diabetes, heart disease, leaky guts, and raging inflammatory processes. It is estimated that almost 80 million Americans are currently prediabetic, with most of them unaware of this fact. If we don't do something soon, our children's generation is slated by researchers to be the first in American history to die younger than their parents.

In order to capitalize on the genomic wisdom that we've inherited and start to reverse this growing trend of unhealthy living, it's time that I introduce you to the 5th and final pillar of the DNA Restart.

It's time that you return to using food as a nurturing and life-sustaining force that supported your ancestors' physical and genomic development over millions of years. The truth is that much of our modern life is devoid of satisfying and nurturing relationships with both people and food. Today our world is one of plenty—of plentiful choices and plentiful supply.

This pillar is about taking everything you've learned so far in the first four pillars and putting it all together in a deeply satisfying, sustainable, and meaningful way. And it's about slowing things down. It's time we return to caring for our most precious gift—our DNA. And we're going to do that together in the 5th and final pillar: Slow Living.

CHAPTER 32

Eating Respectfully

In developing the DNA Restart, I spent quite a lot of time thinking about the health effects of eating speed and mindfulness. This was partly because I knew that if this was a challenging aspect for my adherence to the 5th Pillar, Slow Living, then others might need a little extra help as well.

Research has shown that when we're not in touch with our bodies while we eat, we're going to get obese and even regain weight after losing it. Emotional eating, binge eating, and eating in response to food cravings are often symptoms of being out of touch with our bodies' true needs. Resensitizing ourselves to our natural signals of hunger, fullness, and even taste can help ground us physically and help us eat more in line with what our DNA needs.

In fact, eating mindfully can also help you lose weight and stay thin. But it's going to require some work on your part. Slowing down your eating speed and getting rid of distractions while eating, such as eating incessantly in front of an electronic device or screen, will both go far in getting you to your ideal weight goal.

But most importantly, respectful eating will keep you there.

YOUNG BABIES TEACH SLOW EATING

Babies and young children, who are still in touch with their bodies, naturally practice respectful and intuitive eating every time they encounter food. I'm sure you've witnessed at some point the sight of a baby pushing away food when they are full. That's what we've all lost—that natural connection to our bodies. One of the goals of the DNA Restart is to get your eating back in line with what your body and your DNA actually need, just like babies do every day. As you transition through the next 28 days, you're going to see how your food choices and portions will start to naturally change as you learn to recognize and eat the right foods, and to eat only when you're hungry.

To get you there, I've developed Slow Living Techniques in this 5th Pillar that are based on scientifically proven methods to help you in your weight-loss and maintenance journey. The first of these techniques is called Eating Respectfully, and it's deceptively simple and worth mastering, especially before moving on to having your weekly Slow Meal.

To help you cultivate more respectful eating when you're thinking about or consuming your food on the DNA Restart, you are required to run through the exercise at least once a week.

RESPECTFUL EATING EXERCISE

You'll need 5 minutes without any interruption for this Respectful Eating exercise. You'll also need a single Brazil nut or any other nut from the approved list on page 109; just make sure that it isn't roasted or salted.

Pick up the nut in your hand. Feel its shape and weight. Now smell it. What does it smell like? Does it smell familiar? Does the smell bring back any memories for you? Take your time here and don't rush. Notice that this may actually be the first time in a while that you're in fact smelling your food before "inhaling" or eating it.

Now try gently placing the nut between your teeth, but don't bite in yet. How does the nut feel on your teeth? Hard or soft? How long can you resist before biting in? Now let your teeth cut into the nut and chew very slowly. What's the sensation like as you try to hold the broken pieces between your tongue and palate?

Notice what your tongue is doing while you're chewing. Pay attention now to the sensations that are happening in your mouth. How does the flavor in your mouth change with time? Is there more saliva? How fast are you chewing? Can you slow down your chewing? How long does it take before you feel the urge to swallow?

Now that you've gone through a Respectful Eating exercise, try to see whether you can increase your awareness of the sensations that are happening while you're having your weekly Slow Meal. As you become more aware of the sensations that are happening in your body while you're eating, you will notice that you are getting much more pleasure out of eating and are getting full faster.

I've struggled with overeating for years. Making time to enjoy my food was never a priority. Doing the Respectful Eating exercise wasn't easy for me in the beginning. I kept wanting to swallow the nut! But now, after a few tries, it's made the biggest single difference to get me to stop overeating by getting me to be more respectful when I eat. I try to do it at every meal.

—Celine, 55

CHAPTER 33

Slow Eating the DNA Restart Way

Slow Eating is one of the Slow Living Techniques, which I've developed specifically for you to use in the next 28 days of your DNA Restart and beyond. We all know what it's like to rush through our days and not find the time to sit properly and have a nourishing and relaxing meal. This has only been made worse over the years as we have been spending more of our time eating out (and I'm not speaking about fine dining here) than eating at home.

Because of the breakneck speed in which food is now produced and consumed, modern life has bypassed what was once a natural time of connecting with each other and, most importantly, deriving pleasure from our food: mealtime. Reconnecting to your food, and life in general, is the basis of this 5th and final pillar.

Research supports what we've all instinctively known is true: Eating quickly just isn't good for you. Just one look into your mouth will show you that humans don't have the dental equipment many other animals have to quickly devour their food raw. Compared to other animals, it's not just our teeth that are relatively small. We actually have tiny mouths for our size, which means we evolved to be able to eat densely nutritious food.

One small bite at a time.

Plus, the faster we eat, the more likely we're going to miss the natural satiety cues and end up eating more than we intended or would like to. The result, in a word: obese. This wasn't always the case in our human history.

The lives lived by all of our differing genetic ancestors had them breaking apart nuts between two stones, scraping marrow out of bones with stone tools, and eventually cooking food over fire. As you remember from the 3rd Pillar, Eat Umami, fermenting, curing, and cooking are all amazing ways to boost the amount of deliciousness in our foods. Being able to manipulate our foods in this way not only made them more nutritious, but more delicious as well. And this deliciousness was also an important sign that the food was safe for us to eat.

But all these things require one last important item. And that's time.

It can take more than 15 minutes for your brain and body to start to register the fact that you're getting full. This is why it's so easy to mindlessly eat—and, of course, to catch yourself feeling totally stuffed a little too late. Some of the reason behind the delay is that it takes time for the hormones that are released in your body, such as cholecystokinin, leptin, and insulin, to reach receptors in your body that work to tell you loudly enough that you're full.

Eating is not new for humans, so why the delay? I believe part of the reason for this effect is that the human body, including the brain, evolved to be actively involved in the process of acquiring, preparing, and cooking food, and not just eating it. This is likely because we were all sous chefs to some degree, spending a significant amount of our time processing and cooking our foods very long ago.

So what happens when we slow down that process by becoming more hands-on in preparing a meal?

We eat less.

CHAPTER 34

Food Preparation, Slow Eating, and Weight Loss

I've experienced this effect many times, as you may have as well, whenever I'm involved in preparing a large meal. After hours of chopping, washing, and laboring over a hot stove, when it's finally time to sit down for dinner, I'm just not feeling as hungry as I thought I was. For many years I thought that my sudden surprising feeling of fullness might be a result of my episodic tasting while cooking. But that didn't make much sense, because the amount of food that I was actually eating when tasting for spice levels was insignificant.

To explore this further, I had five of my patients volunteer to take subjective hunger inventories before and after preparing a meal. The only requirements were that they should spend at least 30 minutes doing all of the prep work for the meal, and they were also forbidden from tasting any of the food at all in the kitchen before sitting down to eat their meal.

The results showed me that I wasn't alone. I had observed that the longer I spent in the kitchen prepping, the less I'd eventually eat during the meal. And my patients reported eating less and feeling full much quicker than they would otherwise when they were the ones preparing their meals, even without any tasting or premeal snacking.

The reason for what I discovered about myself, and others, is that we don't eat with only our mouths—but actually with all our senses. Digestion begins by sight and smell alone. Since, for the most part, all our genetic ancestors spent so long acquiring, preparing, and then finally cooking their meals, it should be no surprise that if we let our bodies catch up, we'll end up eating a lot less.

And feeling much better for it.

So let me start by outlining for you what a Slow Eating meal is going to look like. *Your requirement for this part of the 5th Pillar is to have at the bare minimum one Slow Eating meal a week for the next 28 days of your DNA Restart.*

I've even provided a quick prescription and checklist for you to follow to set yourself up for success. The following may sound a lot easier than it actually is, so don't be fooled! Here's what you'll need to do for your once-a-week Slow Meal:

The first and most important thing you're going to need to do for your Slow

Eating meal is to make sure that you're not eating alone, so plan to have this meal with your friends, loved ones, or with other fellow DNA Restarters who live near you.

The second thing you're going to need to do for Slow Eating is to forbid standing. So make sure it's understood that you will have to be dining on a table, floor, picnic blanket, etc. The idea is to force your body into a state of relaxation. So take a load off and get comfortable.

Remember that this is also a time for you to find a real connection—not just with the food you're about to eat but with the other real, three-dimensional, living, breathing human beings with whom you are dining. This is why the third step is to eliminate all of your multitasking temptations for just this one meal. That includes turning off or putting away all electronic devices with a screen. Don't worry; you'll get them back, but at least for this one Slow Eating meal a week, it's really important that you go all-in. This also includes putting away all newspapers, books, and any-thing else that might distract you.

Ambiance plays an enormous role in determining how much we end up eating and how fast. That's why an extra step you can take that will help slow you down naturally and help you to eat less is to dim the lights. Adding a few candles to the table is also a good cue, especially if it's in the evening, that this meal is important. Now that you've taken care of the visual cues, you're also welcome to have some music in the background, as long as it fosters relaxation.

The fourth step is what I call *essential gratitude*. I want you to spend a few min-utes with those around you appreciating the enormity of what you're about to do. Because of the commodification of our dinner tables, we've totally lost touch with the essence of what we consume. You will need to reconnect with the very essence of food so that you can start to hear, feel, and see how it makes you feel. There's only one thing that can bring you the fullness and wholeness you desire, and that's slow-ing down long enough to reconnect with the things that matter most to you.

So use this as an opportunity for gratitude and think of this phase in your Slow Eating exercise as gustatory foreplay. An easy way to transition to doing this is to think about all the things that needed to come together to make your meal possible. For example, consider all the things that the seeds needed, such as sunshine, water, and healthy soil, to germinate and grow. The animals that needed to be fed and cared for. And all of the people who were involved in the process of farming, pick-ing, boxing, and maybe shipping your food. You may never meet them, but they have worked so hard to bring all this bounty to your city, town, neighborhood, and, ultimately, your table.

There's so much that went into the meal you're about to eat, so just take a brief moment to say a simple "thank you" or take some time to fully acknowledge all the

DNA Restart Health Tip #30

Here's your optimal Slow Eating prescription for DNA health and weight loss.

1. Slow Eating means not eating alone.
2. Slow Eating means turning off all screen devices—no texting, dialing, or browsing while eating, whatsoever.
3. Get rid of any other distractions like newspapers and paper books.
4. Take a moment for gratitude.
5. Do the First-Bite/Last-Bite exercise by waiting for 30 seconds before taking your first and last bite of food.

things that had to happen to make the food ingredients that you lovingly prepared almost "magically" appear before you.

Expressing essential gratitude is also good for your DNA. Scientific research is now starting to corroborate that regularly expressing gratitude lowers overall stress levels, which, as you remember from the 2nd Pillar, is good news for your genes. From a digestive perspective, this exercise allows a little time for your body to adjust and start getting ready to eat. You've likely spent most of your day "doing"—now it's time for eating. If there are any young children at the table, a fun exercise to involve them in prior to digging in is to have them try to count up or guess how many and which ingredients went into the food you're about to eat.

Now it's time to eat!

But don't rush it.

To help you in that process, I've developed a simple Slow Living Technique I call the First-Bite/Last-Bite exercise. You're going to try this now, so here's how it works: Get ready to take your first bite, but once you've had your first mouthful of food, wait 30 seconds before taking another bite. By doing this exercise, you're giving yourself a natural pause and allowing your body to really start the process of digestion that will get you feeling full sooner and with having eaten less. Just to give you a heads-up now, you'll be ending the meal in the same way. You should try doing this exercise with every Slow Meal you have. It's also a useful tool to come back to when you feel like you need a little help slowing down.

You should aim to spend at least 30 minutes eating. This way you'll be certain that your body has the time it needs digestively to catch up to your mouth. If you need help extending your eating time, a simple exercise would be to get up and boil some water and prepare your oolong in the middle of the meal. Another thing to keep in mind while you're eating and to discuss with the other DNA Restarters or eating partners present is umami. Is there umami in what you're eating? How does it work with the other flavors present? Is what you're eating delicious? How can you make it even more so? These questions will help refocus your attention back to your body as you're tasting the food.

Now, as you're coming toward the end of what you've intended to eat for your meal, you're going to come back and finish the First-Bite/Last-Bite exercise. So leave at least one mouthful of food on your plate, and wait 30 seconds without eating before finishing it. Then you're free to have your DNA Restart dessert of nuts, a small square of dark chocolate, and/or fruit. By doing this exercise, you're giving yourself a moment to really savor that last bite, just like you did with the first one.

CHAPTER 35

Sleeping Your Way to DNA Health

As we've discussed many times now throughout the first four pillars of the DNA Restart, stress is not only bad for your body, it's terrible for your DNA. That's because stress can lead to a higher natural state of oxidative stress. This was beneficial for all of our various genetic ancestors because the elevated oxidative stress levels helped them fight off infections and parasites much better.

For a contemporary example, think about when you've just been up all night caring for your sick child; your stomach will naturally crank up the amount of hydrochloric acid it produces. This can be beneficial, since more acid can help kill more microbes, like viruses and bacteria that you're now unwittingly being exposed to from your little one. You may notice in sleep-deprived instances like these that your stomach feels a little off. Maybe you'll notice a little more gastric reflux than usual. But regardless of what you consciously notice, your genes notice everything.

Eventually, you somehow manage to get some rest, and your stomach settles back down after the acidic assault. But what happens when you're chronically sleep deprived and stressed? All that extra acid will eventually harm the cells that line your stomach and maybe splash up into your esophagus. Chronic inflammation will also eventually damage the DNA in your cells, and that can lead to cancer. Having chronic inflammation unfortunately has even been associated with a host of debilitating conditions, such as Alzheimer's disease.

Fortunately, as I'm sure you're well aware by now, you're not a slave to these biological processes. And sometimes the simplest interventions are powerfully effective. There is now evidence to suggest, for example, that actively being more mindful or meditating on a regular basis can actually change the way your DNA behaves through epigenetics, and this lowers inflammation levels in the body. The gene *PTGS2* that codes for the *COX-2* enzyme, which we learned about in the 2nd Pillar, was actually found to be turned down low by people practicing mindful meditation.

Like I said, your DNA is always listening.

So how can you reduce oxidative stress, turn off chronic inflammation, and make sure your genes are rested enough so they can do that reparative work they're so good at, all in one go? It's simple—and you don't need to become a master meditator. All you need is sleep.

GIVE YOUR DNA A BREAK

As we've discussed earlier, telomeres are like protective bumpers on the ends of each one of your 46 chromosomes that make up your DNA. As we age, these genetic bumpers naturally shorten with time—which is one sign of your genetic age. Everything from higher chronic inflammation to cardiovascular disease and, of course, obesity has been linked in one way or another with shortened telomeres.

Interestingly, far from any intensive clinical intervention, science is now showing that one very effective way to keep our telomeres long like we like them is to get a solid night's sleep. This emerging research is indicating that if you're not sleeping enough, or the quality of your sleep isn't that great, you may be at risk of having shorter telomeres.

There's really no way around it. Your DNA needs you to sleep. It's not that your DNA needs to rest in the traditional sense, but rather your body needs some quiet and undisturbed time to run repairs, consolidate memories, and ensure that your immune system is working optimally. Studies looking at long-term shift-workers who have a disrupted sleep/wake cycle due to the nature of their work found that study participants have detrimental epigenetic changes in their DNA related to both inflammation and the immune system.

Serious epigenetic consequences aside for a moment—I'm sure you're well aware of what it feels like when you haven't slept enough. Not great. But there are going to be times in your life when that's simply unavoidable.

So what happens to your waistline when you're not getting an optimal amount of sleep? To answer this question, a study conducted in Quebec, Canada, followed families for 6 years to see if they could find a connection between sleep and body weight. What they discovered was that when people either slept too little or even too long, this could accurately predict weight gain even 6 years down the road.

We still do not know for certain whether not sleeping enough just allows us more time to eat or whether it can change the way many of our appetite hormones such as leptin and ghrelin function. What we do know is that if you don't make sure your children get enough sleep, this can have a bigger impact on their levels of belly fat than not exercising enough or watching too much television. Yes, sleep even trumps TV and exercise!

The take-home message here is pretty clear. Why work so hard on the DNA Restart to change how you eat, exercise, and live, without giving your body and genes one of the keys things they need to survive and, ultimately, thrive? We all know we need to get more sleep than we're getting, and now there is solid science to prove it. So what can we possibly do?

There are a few things you should start doing as you commence your 28-day DNA Restart that will not only help you sleep better, but also play a big role in reducing your overall levels of stress. The first is probably the simplest, but one you likely don't give yourself the time for—a warm bath. This is a great and natural way to end a long day and bring your overall level of stimulation down a few very crucial notches. Dimming the lights or using candles and making sure that the water is hot enough can help immeasurably. If you have young children, this might be hard to do given the demands on your time, but remember that the DNA Restart is the time for you to try out the changes you want to make to bring yourself back into healthy alignment with your DNA. Some successful DNA Restarters who have come before you reported that they were in a better mood when putting their kids to bed at night when they anticipated drawing themselves a hot bath right afterward. They even thought of it as a "reward" for all of the positive DNA Restart food choices they had made that day.

If you're going to go to the trouble of drawing yourself a hot bath, there are a few other things you should try as well. I've listed them all in DNA Restart Health Tip #31, and they are all essential to give your body and genes the downtime they need.

What bears highlighting is that it is my medical opinion that you give yourself a hard cutoff for screen time from electronic devices that keep you online. I'm not going to require an exact time because you're all adults, but I would suggest no later than 10:00 p.m.

R̶X̶ *DNA Restart Health Tip #31*

Here's your optimal Slow Sleeping prescription for DNA health and weight loss. Employ it every day.

1. Try having a warm bath after a long day, preferably 1½ hours before bed.

2. Have a sleep routine prepared to employ 30 minutes before bed.

3. Turn down or off all screens in the evening. I suggest a cutoff time of no later than 10:00 p.m.

4. No television, e-mail, or Internet for 30 minutes before bed.

One of the things I've found most helpful is having a sleep-oriented routine. Think of Slow Sleeping as your Slow Meal for your sleep needs. And the more you prepare, the better. This can include putting out your clothes for tomorrow or a nightly skin care regimen—whatever works for you. The goal is to cue your body that it's time to get ready for bed. And don't forget that a waking routine can be just as important as you transition out of your bed and into a new day. It's a good time to thank your body and DNA for all the hard work they were doing while you were sound asleep.

THE DNA RESTART TAKEAWAYS

5th Pillar: Slow Living

1. Remember that Slow Living is the final secret to recalibrating your life to maximize your DNA potential.

2. Do a Respectful Eating exercise at least once a week.

3. Include the First-Bite/Last-Bite exercise at mealtimes at least once a day.

4. Have a Slow Meal at least once a week, remembering to do your daily First-Bite/Last-Bite exercise during the meal itself.

5. Prepare your own foods to program your body to eat less.

6. No screen time for at least 30 minutes before bed or no later than 10:00 p.m.

7. Develop a sleep routine and follow it every night.

8. Give your DNA a break and get more sleep (at least 6 hours and not more than 10 hours per night).

PART VI

THE DNA RESTART
Road Map to Optimal Health
and Longevity

Congratulations! You have now made it all the way through the five pillars of the DNA Restart and have all of the information needed to depart on your successful journey to reaching your ideal weight and optimal genetic health. The final part of this book has been designed to get you to your goals in a step-by-step process. The road map that I've created in this section distills all of the complex, scientifically dense information of the five pillars into easy-to-follow steps so that you can launch into the DNA Restart immediately.

We've covered a lot of information in this book, so I'm providing you with an outline of the major points from each of the five pillars of the DNA Restart to serve as an easy-to-follow road map. Also remember to refer back to the takeaways from each of the five pillars, because some of the exercises (such as the Umami Taste Test Experience and the Cotton Swab Alcohol Intake Test) you'll only need to do once. This will help you keep track of all of the health-promoting steps you'll be taking over the next 28 days.

THE DNA RESTART ROAD MAP
TO OPTIMAL HEALTH AND LONGEVITY

5th Pillar

Do the First-Bite/Last-Bite exercise at least once a day.

Practice Respectful Eating.

Prepare your own meals the DNA Restart way to program your body to eat less and lose weight naturally.

Have a DNA Restart Slow Meal at least once a week.

No screen time for at least 30 minutes before bed and no later than 10:00 p.m.

Develop your own sleep routine and follow it every night.

4th Pillar

Enjoy oolong tea, which is the only polyphenol-rich drink that's naturally calorie-free.

Have 2–4 cups of oolong tea every day, prepared the DNA Restart way.

Unlike green or black tea, oolong tea is easy on the stomach.

Oolong tea protects your DNA from oxidative stress.

Drinking oolong tea daily is naturally slimming, as it will stop you from absorbing fat from your diet.

3rd Pillar

Do the Umami Taste Test Experience.

Enjoy umami-rich protein for breakfast to help you lose weight.

Have two to three servings (2–4 oz. each) of umami-rich seafood a week.

Activate DNA Restart umami synergism by combining different umami foods to get full faster and stay full longer.

2nd Pillar

Activate your innate DNA healing mechanism by exercising 5–6 times a week.

Avoid monoeating.

Have up to four 1-oz. servings of nuts a week.

Drink the juice of a lemon or two limes every day.

Have up to four 2-oz. servings of legumes a week.

Purge all emulsifiers from your diet.

Avoid harmful phytochemicals and mycotoxins.

1st Pillar

Take the DNA Restart Cracker Self-Test to find out your Carb Consumption Category.

Do the DNA Restart Cotton Swab Alcohol Intake Test to determine your weekly alcohol allowance.

Limit your red meat intake (2–3 oz. serving size) to no more than twice a week.

Get your iron levels assessed if you're male or a postmenopausal female.

Ban all processed meats from your diet.

Eat fermented dairy at least once a week if your DNA lets you.

Absolutely no soft drinks or vegetable or fruit juices whatsoever.

No artificial sweeteners.

CHAPTER 36

How to Use the DNA Restart Carb Cost Calculator

W hat you'll need to do now is go back to your results from your genetic self-tests, because you're going to be using them to plan your meals. Below you will find a list of foods and their corresponding carb cost for you to consider when you're deciding what to eat for the week. I've created a DNA Restart Carb Cost Allowance Guide that you can use to keep track of your weekly intake on a points-based system.

Here's a quick reminder of how to calculate your weekly carb allowance:

1. Take the DNA Restart Cracker Self-Test on page 15.

2. Find your Carbohydrate Consumption Category based on the results of your cracker self-test.

3. Your allowance is based upon your Carbohydrate Consumption Category and is how you will "buy" the carbs you'll be eating weekly. Just make sure that you stay within your allowance budget for the week!

4. Even though fruits technically contain carbohydrates, you should enjoy them on the DNA Restart because of their unique, phytochemical-rich profile. Just don't exceed four servings a day, make sure you're choosing from a variety of sources (no monoeating!), and stay far away from any juices.

5. Remember that the carb cost allowance system is meant to be a guide to get you to eat more in line with your DNA as per your results from your self-test and not meant to be a definitive list of all the food that may contain carbohydrates.

Here's the carb point allowance breakdown:

Full Carbs: 13 to 16 points of carbs a week

Moderate Carbs: 9 to 12 points of carbs a week

Restricted Carbs: 5 to 8 points of carbs a week

The main idea behind this section is to provide you with all of the tools that you will need to be successful as you transition through your DNA Restart. Most importantly, remember to enjoy the benefits of your restarted body, genes, and life.

To give you an idea of what your meal plans should look like on the DNA Restart, the section on pages 207 through 218 includes a 2-week sample meal plan and the corresponding *carb cost*, or cc, for each of the Carb Consumption Categories: Full, Moderate, and Restricted. I have also provided a convenient Checklist for Your Personalized DNA Restart Mix and Match Meal Plans (page 219) for you to consult when designing your personalized meal plans for your DNA Restart.

All of the recipes in the DNA Restart list their corresponding carb cost point values to help keep you within your weekly carb cost allowance. The recipes have been specifically developed with the five pillars of the DNA Restart in mind to get you eating delicious umami- and phytonutrient-packed food that will not only help you lose weight but also take care of your precious DNA at the same time.

CHAPTER 37

The DNA Restart Carb Cost Allowance Guide

Breads

Per the 1st Pillar, make sure these are all free of emulsifiers. Refer to page 35 for a list of the various names for emulsifiers.

Bagel—4 points

Bagel, whole wheat—3 points

Bread, white flour (per slice)—
1.5 points

Bread, whole wheat flour
(per slice)—0.5 point

Pita—3 points

Pita, whole wheat—2 points

Rice bread (per slice)—1.5 points

Roll, white flour (small)—3 points

Roll, whole wheat flour (small)—
2 points

Tortilla, white flour (small)—
2 points

Tortilla, whole wheat flour
(small)—1 point

Wrap, white flour (small)—2 points

Wrap, whole wheat (small)—1 point

Cereal Grains/Legumes/Pasta/Rice

Measurements when cooked.

Amaranth, whole grain (½ cup)—
2 points

Chickpeas (½ cup)—1 point

Cream of Wheat (½ cup
prepared)—2 points

Legumes (½ cup)—0.5 point

Oatmeal (½ cup)—0.25 point

Pasta, white flour, any shape
(½ cup)—3 points

Pasta, whole wheat flour, any shape
(½ cup)—1.5 points

Quinoa (½ cup)—1 point

Rice, brown (½ cup)—1 point

Rice, white (½ cup)—3 points

Wild rice (½ cup)—1.5 points

Dairy Products

Kefir (plain, unsweetened)
(1 cup)—0.25 point

Milk, cow (whole, 2%, 1%, and fat-free) (1 cup)—1 point

Yogurt (plain, unsweetened)
(1 cup)—0.5 point

Fruits and Honey

Although both fruits and honey technically have carbs, I'm giving you a pass on this category and making them point-free to make sure you get all the phytonutrients you can. Just make sure that you limit yourself to 2 teaspoons of honey a day and don't overdo your fruit intake (four servings a day). And no juices!

Vegetables

Measurements when cooked.

Carrots (2 medium)—0.5 point

Corn (½ cup)—1 point

Onion (½ cup)—0.25 point

Potato (baked or boiled) (1 cup)—0.5 point

Squash (baked) (½ cup)—1 point

Sweet potato (baked) (½ cup)—0.5 point

CHAPTER 38

The DNA Restart
Sample Weekly Meal Plans*

*Feel free to swap out any of the suggested rice, quinoa, or other whole cereal grains for a vegetable side dish according to your own individual dietary preferences/sensitivities.

Legend:

RED MEAT   CHICKEN   FISH   cc: carbohydrate cost

Full Carbs: Week 1

	BREAKFAST	LUNCH	DINNER
SUNDAY 2–4 cups oolong tea a day	Omega Nut Oatmeal (page 220) cc: 0.25	Spiced Lentil Soup (page 228) cc: 0.5	Easy Baked Miso Fish (page 240) with brown rice cc: 1.0
MONDAY 2–4 cups oolong tea a day	Umami Omelet Bomb (page 223)	Romano Bean, Tomato, Basil, and Mozzarella Salad (page 236) cc: 0.5	Ceylon Cinnamon Beef Stew (page 242) and Golden Saffron Rice (page 230) cc: 3.0
TUESDAY 2–4 cups oolong tea a day	Spiced Pumpkin Oatmeal (page 222) cc: 0.25	Grandma Oolong Eggs (page 232) and Aegean Umami Roasted Tomatoes (page 226)	Saffron Chicken with Vegetable Patch Stew and Almonds (page 241) with brown rice or quinoa cc: 1.25

(continued)

Full Carbs: Week 1 (continued)

	BREAKFAST	LUNCH	DINNER
WEDNESDAY 2–4 cups oolong tea a day	Protein-Powered Oatmeal (page 221) cc: 0.25	Simply Delicious DNA Restart Green Salad (page 234) cc: 0.25	Greek Lamb Chops (page 243) with brown rice or quinoa cc: 1.0
THURSDAY 2–4 cups oolong tea a day	Grandma Oolong Eggs (page 232) and Aegean Umami Roasted Tomatoes (page 226)	Spiced Lentil Soup (page 228) cc: 0.5	Pistachio Sage Encrusted Fish (page 239) with brown rice cc: 1.0
FRIDAY 2–4 cups oolong tea a day	Omega Nut Oatmeal (page 220) cc: 0.25	Aegean Umami Roasted Tomatoes (page 226) with brown rice or quinoa cc: 1.0	Saffron Chicken with Vegetable Patch Stew and Almonds (page 241) with brown rice or quinoa cc: 1.25
SATURDAY 2–4 cups oolong tea a day	Umami Omelet Bomb (page 223) and Ancient Antioxidant Olive Tapenade (page 227)	Vegetable Medley with Walnuts (page 231) and Roasted Spiced Garlic Flowers (page 229) with potatoes cc: 1.0	Easy Baked Miso Fish (page 240) with brown rice or quinoa cc: 1.0
*Total carb cost (cc) allowance used: 14.25 points			

Full Carbs: Week 2

	BREAKFAST	LUNCH	DINNER
SUNDAY 2–4 cups oolong tea a day	Protein-Powered Oatmeal (page 221) cc: 0.25	Simply Delicious DNA Restart Green Salad (page 234) cc: 0.25	Greek Lamb Chops (page 243) with brown rice or quinoa cc: 1.0
MONDAY 2–4 cups oolong tea a day	Omega Nut Oatmeal (page 220) cc: 0.25	Aegean Umami Roasted Tomatoes (page 226) with brown rice or quinoa cc: 1.0	Easy Baked Miso Fish (page 240) with potatoes cc: 0.5
TUESDAY 2–4 cups oolong tea a day	Grandma Oolong Eggs (page 232) and Aegean Umami Roasted Tomatoes (page 226)	Spiced Lentil Soup (page 228) cc: 0.5	Saffron Chicken with Vegetable Patch Stew and Almonds (page 241) with brown rice or quinoa cc: 1.25
WEDNESDAY 2–4 cups oolong tea a day	Umami Omelet Bomb (page 223)	Vegetable Medley with Walnuts (page 231) with brown rice or quinoa cc: 1.5	Ceylon Cinnamon Beef Stew (page 242) with Golden Saffron Rice (page 230) cc: 3.0

(continued)

Full Carbs: Week 2 *(continued)*

	BREAKFAST	LUNCH	DINNER
THURSDAY 2–4 cups oolong tea a day	Omega Nut Oatmeal (page 220) cc: 0.25	Grandma Oolong Eggs (page 232) and Aegean Umami Roasted Tomatoes (page 226)	Pistachio Sage Encrusted Fish (page 239) with brown rice or quinoa cc: 1.0
FRIDAY 2–4 cups oolong tea a day	Spiced Pumpkin Oatmeal (page 222) cc: 0.25	Spiced Lentil Soup (page 228) cc: 0.5	Saffron Chicken with Vegetable Patch Stew and Almonds (page 241) with brown rice or quinoa cc: 1.25
SATURDAY 2–4 cups oolong tea a day	Umami Omelet Bomb (page 223) and Ancient Antioxidant Olive Tapenade (page 227)	Romano Bean, Tomato, Basil, and Mozzarella Salad (page 236) cc: 0.5	Easy Baked Miso Fish (page 240) with potatoes cc: 0.5
*Total carb cost (cc) allowance used: 13.75 points			

Moderate Carbs: Week 1

	BREAKFAST	LUNCH	DINNER
SUNDAY 2–4 cups oolong tea a day	Omega Nut Oatmeal (page 220) cc: 0.25	Simply Delicious DNA Restart Green Salad (page 234) cc: 0.25	Greek Lamb Chops (page 243) with brown rice or quinoa cc: 1.0
MONDAY 2–4 cups oolong tea a day	Umami Omelet Bomb (page 223)	Romano Bean, Tomato, Basil, and Mozzarella Salad (page 236) cc: 0.5	Saffron Chicken with Vegetable Patch Stew and Almonds (page 241) and DNA Restart Rosemary Mashed Potatoes (page 237) cc: 0.75
TUESDAY 2–4 cups oolong tea a day	Protein-Powered Oatmeal (page 221) cc: 0.25	Vegetable Medley with Walnuts (page 231) and Aegean Umami Roasted Tomatoes (page 226) cc: 0.5	Easy Baked Miso Fish (page 240) with potatoes cc: 0.5
WEDNESDAY 2–4 cups oolong tea a day	Grandma Oolong Eggs (page 232) and Aegean Umami Roasted Tomatoes (page 226)	Spiced Lentil Soup (page 228) cc: 0.5	Saffron Chicken with Vegetable Patch Stew and Almonds (page 241) with brown rice or quinoa cc: 1.25

(continued)

Moderate Carbs: Week 1 *(continued)*

	BREAKFAST	LUNCH	DINNER
THURSDAY 2–4 cups oolong tea a day	Spiced Pumpkin Oatmeal (page 222) cc: 0.25	Grandma Oolong Eggs (page 232) and Aegean Umami Roasted Tomatoes (page 226)	Easy Baked Miso Fish (page 240) with brown rice or quinoa cc: 1.0
FRIDAY 2–4 cups oolong tea a day	Umami Yogurt with Berries (page 224) cc: 0.25	Spiced Lentil Soup (page 228) cc: 0.5	Ceylon Cinnamon Beef Stew (page 242) and Golden Saffron Rice (page 230) cc: 3.0
SATURDAY 2–4 cups oolong tea a day	Umami Omelet Bomb (page 223) and Ancient Antioxidant Olive Tapenade (page 227)	Vegetable Medley with Walnuts (page 231) and Aegean Umami Roasted Tomatoes (page 226) cc: 0.5	Easy Baked Miso Fish (page 240) with potatoes cc: 0.5
*Total carb cost (cc) allowance used: 11.75 points			

Moderate Carbs: Week 2

	BREAKFAST	LUNCH	DINNER
SUNDAY 2–4 cups oolong tea a day	Umami Yogurt with Berries (page 224) cc: 0.25	Spiced Lentil Soup (page 228) cc: 0.5	Greek Lamb Chops (page 243) with brown rice or quinoa cc: 1.0
MONDAY 2–4 cups oolong tea a day	Omega Nut Oatmeal (page 220) cc: 0.25	Vegetable Medley with Walnuts (page 231) with brown rice or quinoa cc: 1.5	Pistachio Sage Encrusted Fish (page 239) and DNA Restart Rosemary Mashed Potatoes (page 237) cc: 0.25
TUESDAY 2–4 cups oolong tea a day	Spiced Pumpkin Oatmeal (page 222) cc: 0.25	Vegetable Medley with Walnuts (page 231) and Aegean Umami Roasted Tomatoes (page 226) cc: 0.5	Saffron Chicken with Vegetable Patch Stew and Almonds (page 241) with brown rice or quinoa cc: 1.25
WEDNESDAY 2–4 cups oolong tea a day	Protein-Powered Oatmeal (page 221) cc: 0.25	Simply Delicious DNA Restart Green Salad (page 234) cc: 0.25	Ceylon Cinnamon Beef Stew (page 242) with brown rice or quinoa cc: 1.25

(continued)

Moderate Carbs: Week 2 (continued)

	BREAKFAST	LUNCH	DINNER
THURSDAY 2–4 cups oolong tea a day	Umami Omelet Bomb (page 223) with Ancient Antioxidant Olive Tapenade (page 227)	Romano Bean, Tomato, Basil, and Mozzarella Salad (page 236) cc: 0.5	Easy Baked Miso Fish (page 240) with potatoes cc: 0.5
FRIDAY 2–4 cups oolong tea a day	Spiced Pumpkin Oatmeal (page 222) cc: 0.25	Fresh Greens Salad with Herbs and Spiced Nuts (page 235)	Saffron Chicken with Vegetable Patch Stew and Almonds (page 241) with brown rice or quinoa cc: 1.25
SATURDAY 2–4 cups oolong tea a day	Grandma Oolong Eggs (page 232) and Aegean Umami Roasted Tomatoes (page 226)	Spiced Lentil Soup (page 228) cc: 0.5	Pistachio Sage Encrusted Fish (page 239) with brown rice cc: 1.0
*Total carb cost (cc) allowance used: 11.50 points			

Restricted Carbs: Week 1

	BREAKFAST	LUNCH	DINNER
SUNDAY 2–4 cups oolong tea a day	Umami Yogurt with Berries (page 224) cc: 0.25	Spiced Lentil Soup (page 228) cc: 0.5	Greek Lamb Chops (page 243) with brown rice or quinoa cc: 1.0
MONDAY 2–4 cups oolong tea a day	Umami Omelet Bomb (page 223) with Ancient Antioxidant Olive Tapenade (page 227)	Vegetable Medley with Walnuts (page 231) and Aegean Umami Roasted Tomatoes (page 226) cc: 0.5	Easy Baked Miso Fish (page 240) and Simply Delicious DNA Restart Green Salad (page 234) cc: 0.25
TUESDAY 2–4 cups oolong tea a day	Protein-Powered Oatmeal (page 221) cc: 0.25	Fresh Greens Salad with Herbs and Spiced Nuts (page 235)	Saffron Chicken with Vegetable Patch Stew and Almonds (page 241) and DNA Restart Rosemary Mashed Potatoes (page 237) cc: 0.75
WEDNESDAY 2–4 cups oolong tea a day	Grandma Oolong Eggs (page 232) and Aegean Umami Roasted Tomatoes (page 226)	Spiced Lentil Soup (page 228) cc: 0.5	Pistachio Sage Encrusted Fish (page 239) with brown rice or quinoa cc: 1.0

(continued)

Restricted Carbs: Week 1 *(continued)*

	BREAKFAST	LUNCH	DINNER
THURSDAY 2–4 cups oolong tea a day	Umami Yogurt with Berries (page 224) cc: 0.25	Umami Omelet Bomb (page 223) and Aegean Umami Roasted Tomatoes (page 226)	Ceylon Cinnamon Beef Stew (page 242) and Vegetable Medley with Walnuts (page 231) cc: 0.75
FRIDAY 2–4 cups oolong tea a day	Protein-Powered Oatmeal (page 221) cc: 0.25	Romano Bean, Tomato, Basil, and Mozzarella Salad (page 236) cc: 0.5	Easy Baked Miso Fish (page 240) and DNA Restart Rosemary Mashed Potatoes (page 237) cc: 0.25
SATURDAY 2–4 cups oolong tea a day	Umami Yogurt with Berries (page 224) cc: 0.25	Vegetable Medley with Walnuts (page 231) and Aegean Umami Roasted Tomatoes (page 226) cc: 0.5	Saffron Chicken with Vegetable Patch Stew and Almonds (page 241) cc: 0.25
*Total carb cost (cc) allowance used: 8 points			

Restricted Carbs: Week 2

	BREAKFAST	LUNCH	DINNER
SUNDAY 2–4 cups oolong tea a day	Umami Omelet Bomb (page 223) with Ancient Antioxidant Olive Tapenade (page 227)	Vegetable Medley with Walnuts (page 231) and Aegean Umami Roasted Tomatoes (page 226) cc: 0.5	Greek Lamb Chops (page 243) with brown rice or quinoa cc: 1.0
MONDAY 2–4 cups oolong tea a day	Protein-Powered Oatmeal (page 221) cc: 0.25	Spiced Lentil Soup (page 228) cc: 0.5	Easy Baked Miso Fish (page 240) with potatoes cc: 0.5
TUESDAY 2–4 cups oolong tea a day	Grandma Oolong Eggs (page 232) and Aegean Umami Roasted Tomatoes (page 226)	Romano Bean, Tomato, Basil, and Mozzarella Salad (page 236) cc: 0.5	Saffron Chicken with Vegetable Patch Stew and Almonds (page 241) and DNA Restart Rosemary Mashed Potatoes (page 237) cc: 0.75
WEDNESDAY 2–4 cups oolong tea a day	Protein-Powered Oatmeal (page 221) cc: 0.25	Fresh Greens Salad with Herbs and Spiced Nuts (page 235)	Ceylon Cinnamon Beef Stew (page 242) and Vegetable Medley with Walnuts (page 231) cc: 0.75

(continued)

Restricted Carbs: Week 2 *(continued)*

	BREAKFAST	LUNCH	DINNER
THURSDAY 2–4 cups oolong tea a day	Spiced Pumpkin Oatmeal (page 222) cc: 0.25	Vegetable Medley with Walnuts (page 231) cc: 0.5	Saffron Chicken with Vegetable Patch Stew and Almonds (page 241) cc: 0.25
FRIDAY 2–4 cups oolong tea a day	Umami Yogurt with Berries (page 224) cc: 0.25	Umami Omelet Bomb (page 223) and Aegean Umami Roasted Tomatoes (page 226)	Easy Baked Miso Fish (page 240) and Vegetable Medley with Walnuts (page 231) cc: 0.5
SATURDAY 2–4 cups oolong tea a day	Protein-Powered Oatmeal (page 221) cc: 0.25	Spiced Lentil Soup (page 228) cc: 0.5	Pistachio Sage Encrusted Fish (page 239) and Fresh Greens Salad with Herbs and Spiced Nuts (page 235)
*Total carb cost (cc) allowance used: 7.5 points			

Checklist for Your Personalized
DNA Restart Mix and Match Meal Plans

☐ Determine your DNA Restart Carbohydrate Consumption Category (page 16).

☐ When planning your weekly meals, keep in mind your personalized carb cost allowance (consult the recipes for carb cost per serving).

☐ Eat fermented dairy products at least once a week if your genes allow.

☐ Have umami-rich protein for breakfast to lose weight faster.

☐ Drink 2 to 4 cups of oolong tea every day, preferably with meals.

☐ Limit your coffee intake to 2 cups a day (with 1% or 2% milk, no cream or sugar).

☐ Have up to four servings (1 ounce each) of nuts a week from the approved list (see page 109). If you're planning on doing the DNA Restart as a vegan or vegetarian, you can increase your weekly nut intake up to a maximum of 8 servings (1 ounce each).

☐ Have dessert after every dinner; this could be a serving of fruits and nuts or dark chocolate (1 ounce of at least 72% cacao twice a week).

☐ If you're choosing to consume alcohol on the DNA Restart, then adhere to your optimal weekly alcohol allowance (see page 51) based on the results of your genetic self-test (see page 50), and remember to always drink your alcohol with your meals.

☐ Have no more than two servings (2 to 3 ounces each) of red meat a week.

☐ Have two to three servings (2 to 4 ounces each) per week of umami-rich seafood from the DNA Restart approved list (see page 102).

☐ Combine umami foods for weight loss using the Synergism of Umami Taste table (page 161).

☐ Eat at least three or four olives twice a week or a serving of the Ancient Antioxidant Olive Tapenade (page 227) once a week.

☐ Have up to 4 servings (about 2 ounces each) of legumes a week.

☐ Eat no more than seven or eight whole eggs (preferably pastured) a week.

☐ Plan to have a Slow Meal once a week and do the First-Bite/Last-Bite exercise daily.

CHAPTER 39

The DNA Restart Recipes and Food Tips

Legend

(DF)—Dairy-Free

(GF)—Gluten-Free

(VG)—Vegetarian

(VN)—Vegan

(VN*)—Vegan Adaptable

Many of the recipes are adaptable for vegetarians and vegans, and options have been provided wherever possible.

Please note: When preparing the following recipes, be sure to exclude all allergens according to your individual sensitivities.

Omega Nut Oatmeal

(DF) (GF) (VN*)   MAKES 2 SERVINGS (0.25 cc per serving)

½ cup instant gluten-free rolled oats

1–1½ cups boiling water

1 tablespoon ground flaxseed

½ teaspoon unpasteurized honey*

½ banana, sliced

Pinch of ground Ceylon cinnamon

1 teaspoon raw pistachios, shelled with skins intact

1 teaspoon raw walnuts, shelled with skins intact

1. In a small bowl, combine the oats and ½ cup of the boiling water. Add more boiling water as needed to reach your desired consistency.

2. Add the flaxseed, honey, banana, and cinnamon and mix thoroughly. Top with the pistachios and walnuts.

If you'd like to prepare this recipe to be vegan-friendly, leave the honey out.

DNA Restart Food Tip: As you might remember from the 2nd Pillar, Reverse Aging, eating certain tree nuts—like the pistachios and walnuts in this recipe—is a powerful way to protect your DNA.

Protein-Powered Oatmeal

(GF) (VN*)   MAKES 2 SERVINGS (0.25 cc per serving)

½ cup instant gluten-free rolled oats

1–1½ cups boiling water

½ teaspoon unpasteurized honey*

½ banana, sliced

¼ teaspoon unsweetened nonalkaline cocoa powder

2 ounces organic grass-fed whey concentrate*

1. In a small bowl, combine the oats and ½ cup of the boiling water. Add more boiling water as needed to reach your desired consistency.

2. Add the honey, banana, and cocoa and mix thoroughly. Add the whey concentrate and stir to combine, adding additional water if desired to thin the mixture.

*If you'd like to prepare this recipe to be vegan-friendly, switch the whey concentrate to a powder sourced from plant protein, such as pea.

DNA Restart Food Tip: As you might remember from the 3rd Pillar, Eat Umami, research has shown that starting your day with a protein-packed, umami-rich breakfast is the best way to lose weight. This protein-packed oatmeal recipe is a delicious reminder of why you should never skip breakfast!

"No-Fail" Quinoa

(DF) (GF) (VN)   MAKES 4 SERVINGS (1 cc per serving)

1¼ cups water

1 cup dry white, red, or black quinoa

¼ teaspoon sea salt

In a medium saucepan over medium heat, combine the water, quinoa, and salt. Cover and bring to a boil. Reduce the heat to low and cook, covered, for 20 minutes, or until the water is absorbed. Remove from the heat and allow to rest, covered, for 10 minutes. Remove the lid, fluff with a fork, and serve.

Spiced Pumpkin Oatmeal

(DF) (GF) (VN*)   MAKES 2 SERVINGS (0.25 cc per serving)

½ cup instant gluten-free rolled oats

1–1½ cups boiling water

¼ cup canned unsweetened organic pureed pumpkin

½ teaspoon unpasteurized honey*

Pinch of ground allspice

Pinch of ground Ceylon cinnamon

Pinch of ground cloves

Pinch of ground ginger

Pinch of ground nutmeg

1. In a small bowl, combine the oats and ½ cup of the boiling water. Add more boiling water as needed to reach your desired consistency.

2. Add the pumpkin and honey to the oats and mix to combine. Sprinkle the allspice, cinnamon, cloves, ginger, and nutmeg over the oats and stir well. Add additional water if desired to thin the mixture.

*If you'd like to prepare this recipe to be vegan-friendly, leave the honey out.

DNA Restart Food Tip: As you might remember from the 2nd Pillar, Reverse Aging, research has shown that spices like cloves, ginger, and nutmeg are all great sources of antioxidants. There's no better way to protect your DNA from the get-go in the a.m.!

Umami Omelet Bomb

(DF) (VG)   MAKES 2 SERVINGS

3 tablespoons dried mushrooms
(shiitake or porcini)

⅓ cup boiling water

4 eggs

1 pea-size dollop anchovy paste

2 drops Worcestershire sauce

1 tablespoon water

1 teaspoon coconut or MCT oil

1 tablespoon Parmesan cheese

Freshly ground pepper to taste

1. Place the mushrooms in a measuring cup and pour the boiling water over them. Set aside for 5 minutes. Meanwhile, crack the eggs into a bowl and whisk until the yolks are incorporated.

2. Add the anchovy paste to the mushroom mixture and stir to dissolve. Stir in the Worcestershire sauce. Pour the mixture into the whisked eggs and stir to combine.

3. In a medium skillet over medium heat, heat the 1 tablespoon water and the oil. Pour the egg mixture into the pan and cook until set. Sprinkle half the omelet with the cheese. Fold the omelet in half, turn over, and cook until done. Season to taste with the pepper.

DNA Restart Food Tip: As you might remember from the 3rd Pillar, Eat Umami, research has shown that umami helps to make us feel full. This umami bomb is packed full of four umami sources—mushrooms, anchovy paste, Worcestershire sauce, and Parmesan cheese. This synergistic umami will keep you full for hours.

Umami Yogurt with Berries

(GF) (VG)   MAKES 2 SERVINGS (0.25 cc per serving)

1 cup yogurt (plain, unsweetened,
full fat or reduced fat, or Greek)

2 ounces organic grass-fed whey
concentrate

½ teaspoon MCT oil

2 tablespoons Spiced Mixed Nuts
(page 233)

¼ cup berries (seasonal options
preferred)

In a small bowl, combine the yogurt and whey concentrate and stir until smooth.
Add the oil and stir to combine. Sprinkle the nuts and berries over the top.

DNA Restart Food Tip: Adding MCT oil to your yogurt breakfast bowl like in this
recipe is a great way to deliver the fat-soluble phytonutrients contained in the
berries and spiced nuts straight into your body by increasing their absorption
due to the oil.

Easy Sprouted Legumes

(DF) (GF) (VN)   MAKES 4 SERVINGS (0.25 cc per serving)

1 cup any whole legume (such as
chickpeas or lentils)

1. Place the legumes in a colander and briefly rinse under running water. Transfer to a bowl or measuring cup. Add water until the legumes are submerged. Allow to soak overnight at room temperature.

2. In the morning, transfer the legumes to a colander and rinse under running water. Drain. Line a tray or dish with 2 paper towels. Transfer the legumes to the tray or dish. (The tray or dish should be large enough that the legumes aren't any more than 2 to 3 deep.)

3. Once a day, check on the legumes. If they are dry to the touch, moisten a paper towel and apply it on top of the legumes for 5 minutes. Then remove and discard the towel. Avoid getting the legumes too wet, because this can promote mold growth. Depending upon the ambient temperature and the legumes themselves, you should see sprouting in 3 or 4 days. Your legumes are now ready to harvest.

DNA Restart Food Tip: As you might remember from the 2nd Pillar, Reverse Aging, legumes are a great way to take care of your DNA. But to do that efficiently, you need to reduce the phytate levels, and the best way to do that is to sprout your legumes. Remember that having some phytates is okay, as they bind non-DNA-friendly heavy metals like cadmium and arsenic.

Aegean Umami Roasted Tomatoes

(DF) (GF) (VN)   MAKES 2 SERVINGS

1 clove garlic

4 medium tomatoes

2 teaspoons extra-virgin olive oil,
    divided

1 teaspoon balsamic vinegar

2 tablespoons chopped fresh oregano

1. Crush the clove of garlic and then chop it finely. Set it aside for at least 5 minutes before using.

2. Preheat the oven to 250°F. Line a baking sheet with bleach-free parchment paper.

3. Cut the tomatoes in half and arrange them on the baking sheet. Drizzle with 1 teaspoon of the olive oil and the vinegar. Sprinkle some of the chopped garlic and oregano over each tomato half.

4. Bake for 1 hour, or until soft. Allow to cool for 10 minutes before serving. When cooled, sprinkle with the remaining 1 teaspoon olive oil.

DNA Restart Food Tip: As you might remember from the 3rd Pillar, Eat Umami, roasting tomatoes with their seeds intact helps to increase their umami quotient significantly and leads you to get satiated faster.

Ancient Antioxidant Olive Tapenade

(DF) (GF) (VN*)   MAKES 6–8 SERVINGS

1 cup brined kalamata olives (with pits)	1 tablespoon extra-virgin olive oil
½ clove garlic	1½ tablespoons fresh lemon juice
2 small anchovies or 1 pea-size dollop anchovy paste*	Pinch of cumin

1. To remove as much salt as possible from the brining process, place the olives in a large bowl and cover with water. Transfer the olives to a colander and rinse for 30 seconds under running water. Slightly agitate the colander as you rinse to make sure that the olives are thoroughly rinsed.

2. For your DNA to get the full benefit of this tapenade, you will need to use olives with pits. There are two ways you can proceed here. The first method is to use a cherry/olive pitter, but I've found that this technique results in the loss of a lot of olive flesh. The other method you can try is to place the olives between 2 towels (they're going to get soiled) with a cutting board underneath. Now you bang away using either the bottom of a heavy pan or a hammer on the covered olives. Remove the top towel and discard the pits.

3. Crush the garlic and set it aside for 5 minutes before using. In a food processor, combine the pitted olives and the anchovies or anchovy paste. Process until smooth. Add the olive oil and process to combine. Add the lemon juice, cumin, and reserved garlic. Process until combined. Let the tapenade rest for 1 hour before serving to allow the flavors to blend.

If you'd like to prepare this recipe to be vegan-friendly, simply leave out the anchovies or anchovy paste.

DNA Restart Food Tip: As you might remember from the 2nd Pillar, Reverse Aging, olives and their oil are a great way to take care of your DNA. Unfortunately, up to 80 percent of olives' phytonutrients can be lost in the commercial pitting process. That's why I suggest that you pit the olives in this recipe by hand. Your DNA will love you for it!

Spiced Lentil Soup

(DF) (GF) (VN*)   MAKES 4 SERVINGS (0.5 cc per serving)

2 cloves garlic

1 tablespoon extra-virgin olive oil

2 onions, chopped

1 cup sprouted lentils (page 225)

1 small anchovy or 1 pea-size dollop anchovy paste*

2 tablespoons double-concentrate BPA-free tomato paste

3 tablespoons fresh lemon juice, divided

½ teaspoon ground cumin

4 cups water

Sea salt and freshly ground black pepper to taste

3–5 sprigs Italian parsley

1. Crush the garlic, chop finely, and set aside.

2. In a pot over medium heat, heat 1 tablespoon of water and the olive oil. Cook the onions, stirring frequently, for 3 to 4 minutes, or until translucent. Add the sprouted lentils, anchovy or anchovy paste, tomato paste, 1½ tablespoons of the lemon juice, the cumin, water, and the reserved garlic. Increase the heat to high and bring to a boil. Reduce the heat and simmer for 40 minutes, or until the lentils appear to have burst open.

3. Taste the soup to determine if the lentils have been thoroughly cooked. Once they have, turn off the heat, add the remaining 1½ tablespoons lemon juice, and salt and pepper to taste. Garnish with the parsley before serving.

*If you'd like to prepare this recipe to be vegan-friendly, leave out the anchovy or anchovy paste.

DNA Restart Food Tip: Scientific research has now confirmed that the consumption of legumes reduces both cardiovascular and coronary heart disease, and even lowers cholesterol. To make digestion easier for you, blend the soup with a hand blender once it has cooled, which has been shown to reduce instances of passing the dreaded wind.

Roasted Spiced Garlic Flowers

(DF) (GF) (VN)   MAKES 6 SERVINGS

3 heads garlic

1 tablespoon extra-virgin olive oil

3 pinches of ground cumin*

3 pinches of ground Ceylon cinnamon

3 pinches of paprika

Sea salt and freshly ground black pepper to taste

1. With a large knife, cut approximately ¾" to 1" off the top of each of the heads of garlic. Set aside for 5 minutes.

2. Preheat the oven to 350°F. Line a baking sheet with bleach-free parchment paper and place the heads of garlic on the paper, cut side facing up. Drizzle the olive oil evenly over the cut surface of each garlic head. Sprinkle a pinch of the cumin, cinnamon, and paprika over each of the garlic heads. Season with the salt and black pepper.

3. Bake for 15 to 20 minutes, or until soft when squeezed. Let cool for 15 minutes before serving.

Adding spices is optional, but highly recommended!

DNA Restart Food Tip: Remember from the 2nd Pillar, Reverse Aging, that the phytochemical in garlic called alliin needs to be converted by an enzyme called alliinase into allicin, but the enzyme is heat sensitive. To maximize the health benefits from your garlic, always crush or cut it first and wait at least 5 minutes before using.

Golden Saffron Rice

(DF) (GF) (VN)   MAKES 4 SERVINGS (3 cc per serving)

Pinch or 25 threads of saffron   2 leeks
2 cups boiling water   1 tablespoon MCT oil
½–¾ teaspoon salt   1 cup Thai jasmine rice

1. Place the saffron in a large bowl. Pour the boiling water over it and allow to infuse for 1 hour. Stir in the salt.

2. Meanwhile, thoroughly wash the leeks and remove and discard the hard, fibrous green tips. Cut the leeks in half. Rinse under cold water to ensure all of the dirt and grit has been thoroughly removed, checking in-between the layers carefully. Coarsely slice.

3. In a large saucepan over medium heat, heat the oil. Cook the leeks for 3 minutes, or until soft and translucent. Add the rice and cook, stirring constantly, for 1 minute, or until the rice is toasted. Add the saffron-infused water, reduce the heat to low, and cover the saucepan. Simmer for 20 minutes, or until the water is absorbed. Allow the rice to rest for 20 minutes before serving.

Vegetable Medley with Walnuts

(DF) (GF) (VN)   MAKES 4 SERVINGS (0.5 cc per serving)

3 cloves garlic

1–2 teaspoons fresh oregano

2 red bell peppers

2 medium carrots

1 medium red onion

1 medium zucchini

3 tablespoons shelled walnuts with skins intact

1 tablespoon extra-virgin olive oil

Pinch of sea salt

1. Preheat the oven to 250°F. Line a baking sheet with bleach-free parchment paper. Crush and then coarsely chop the garlic cloves. Set aside for at least 5 minutes.

2. Meanwhile, coarsely chop the oregano. Halve the peppers lengthwise (but don't deseed them, unless you have a condition like diverticulosis) and then cut each length in half. Peel and cut the carrots into ½" pieces. Cut the onion crosswise into ½"-thick circles. Slice the zucchini lengthwise and cut into thirds.

3. Place the vegetables, walnuts, and reserved garlic on the baking sheet. Drizzle with the olive oil and toss to coat. Sprinkle with the oregano and salt. Toss to coat.

4. Increase the oven temperature to 300°F and roast the vegetables for 1 to 2 hours, or until tender. Let stand for 10 minutes before eating.

DNA Restart Food Tip: Walnuts have phytonutrients that are good for your DNA because they can stop the damaging cascade of transcribing and translating inflammatory genetic signals that leads to genetic aging.

Grandma Oolong Eggs

(DF) (GF) (VG)   MAKES 3 SERVINGS

6 eggs

1–2 oolong tea bags or 0.5 ounce
loose-leaf oolong tea

½ teaspoon sea salt

3 tablespoons tamari

2 pieces star anise

1 stick cinnamon

1 teaspoon cracked black
peppercorns

1. Place the eggs in a pot and cover with water until the eggs are halfway submerged. Add the oolong and salt. Bring to a boil over high heat. Boil for 8 minutes.

2. Carefully transfer the eggs to a bowl and crack their shells using the back of a spoon. Add the tamari, star anise, cinnamon, and peppercorns to the cooking water. Return the eggs to the pot, reduce the heat to low, and simmer for 2 hours. Remove the eggs and allow them to cool before peeling.

DNA Restart Food Tip: Eggs are an incredible source of choline, with 99 percent found exclusively in the yolk. Choline is used to make neurotransmitters in the brain, and a deficiency puts you at risk for cancer. It also plays a pivotal role in protecting your DNA strands from physically breaking apart. Males need much more choline than females—so men, don't take those yolks out of your omelets!

Spiced Mixed Nuts

(DF) (GF) (VG)   MAKES 4 SERVINGS

¼ cup shelled pistachios, with skins intact

¼ cup shelled walnuts, with skins intact

¼ cup shelled almonds, with skins intact

¼ cup shelled pecans, with skins intact

¼ teaspoon ground Ceylon cinnamon

¼ teaspoon ground cumin

¼ teaspoon ground turmeric

4 teaspoons honey

1. Preheat the oven to 275°F. Line a baking sheet with bleach-free parchment paper.

2. In a medium bowl, combine the nuts. Add the cinnamon, cumin, and turmeric and stir to combine. Drizzle the honey over the mixture and stir to coat the nuts evenly. Transfer the mixture to the baking sheet, spreading the nuts in a single layer. Bake for 20 to 25 minutes, or until the nuts darken and smell toasted. Allow to cool for 15 minutes before serving.

DNA Restart Food Tip: Tree nuts are one of the richest sources of antioxidant phenolics; only certain spices and fruits contain more. Chestnuts, pecans, pistachios, and walnuts are especially good sources.

Simply Delicious DNA Restart Green Salad

(DF) (GF) (VN)   MAKES 2 SERVINGS

1 head butter lettuce, washed, spun, and torn by hand into bite-size pieces

3 rainbow-colored carrots, peeled and sliced into ribbons using peeler

½ pint cherry tomatoes, halved

¼ cup macadamia nuts, raw or roasted, finely chopped

2 spring onions, finely chopped (optional)

3 tablespoons fresh lemon juice

2 tablespoons extra-virgin olive oil

Sea salt to taste

In a large bowl, combine the lettuce, carrots, tomatoes, nuts, and onions. Drizzle with the lemon juice and olive oil, and season to taste with the salt. Gently toss to coat the vegetables. Serve immediately.

Fresh Greens Salad
with Herbs and Spiced Nuts

(DF) (GF) (VN)   MAKES 2 SERVINGS

Dressing

3 tablespoons fresh lemon juice

1 tablespoon walnut oil

1 tablespoon pistachio oil

2 tablespoons extra-virgin olive oil

Salad

1 cup chicory

1 cup romaine

1 cup dandelion

1 cup spinach

1 small bunch parsley (stems removed)

1 small bunch dill (stems removed)

1 medium tomato

Sea salt and freshly ground black pepper to taste

¼ cup Spiced Mixed Nuts (page 233)

1. *To make the dressing:* In a medium cup or small bowl, combine the lemon juice, walnut oil, pistachio oil, and olive oil. Whisk until blended.

2. *To make the salad:* Wash and tear by hand the greens and herbs and place in a salad bowl. Cut the tomato into 8 wedges and add to the bowl. Pour the dressing over the salad and season to taste with the salt and pepper. Toss to coat evenly. Sprinkle the spiced nuts over the top.

Romano Bean, Tomato, Basil, and Mozzarella Salad

(GF) (VG)   MAKES 4 SERVINGS (0.5 cc per serving)

1 can organic Romano beans, rinsed and drained

2 ripe tomatoes, chopped into large chunks

1 bunch fresh basil, leaves torn by hand into bite-size pieces

1 pound fresh bocconcini

¼ cup extra-virgin olive oil

3 tablespoons fresh lemon juice

Fine sea salt to taste

In a medium bowl, combine the beans, tomatoes, basil, and bocconcini. Drizzle with the olive oil and lemon juice, and season to taste with the salt. Toss to coat. Serve immediately.

DNA Restart Rosemary Mashed Potatoes

(DF) (GF) (VN)   MAKES 4 SERVINGS (0.25 cc per serving)

1 pound assorted colored potatoes
  (Yukon Gold, purple, and/or red
  potatoes), skin on
2 teaspoons sea salt, divided

1 teaspoon MCT oil
1 teaspoon extra-virgin olive oil
1 tablespoon finely chopped
  fresh rosemary

1. Wash and gently scrub the potatoes, leaving the skin intact. Cut into uniform, medium chunks. Place in a large pot, cover with cold water, and add 1 teaspoon of the salt. Cover, bring to a gentle boil over medium-low heat, and cook for 20 minutes, or until a fork easily pierces through the flesh of a potato.

2. Drain the potatoes and transfer to a large bowl. Add the MCT oil, olive oil, rosemary, and remaining 1 teaspoon salt. Using a potato masher, mash until the potatoes reach the desired consistency. Taste and adjust seasoning as desired.

Antioxidant Protein Marinade

(DF) (GF) (VN)   MAKES APPROXIMATELY ½ CUP

¼ cup red wine   ½ teaspoon dried oregano
¼ cup fresh lemon juice   ½ teaspoon dried thyme
2 tablespoons extra-virgin olive oil   ¼ teaspoon sea salt

In a mason jar, combine the wine, lemon juice, and olive oil and shake well. Add the oregano, thyme, and salt and shake to combine. This marinade can keep for 2 to 3 days in the refrigerator, but it's best when used immediately.

DNA Restart Food Tip: You can use this marinade whenever you're cooking with a high-protein ingredient. Research has shown that marinating with wine prior to cooking can reduce certain types of heterocyclic amines by up to 88 percent, and lemon juice has been shown to reduce the amount of advanced glycation end-products that are produced during cooking.

Pistachio Sage Encrusted Fish

(DF) (GF)   MAKES 4 SERVINGS

4 fresh or frozen fish fillets (4 ounces each) (see DNA Restart fish choices on page 102)

3 tablespoons fresh lemon juice

½ cup shelled raw pistachios, finely chopped

2 tablespoons chopped fresh sage

2 tablespoons grated Parmesan cheese (aged at least 18 months)

Grated peel of 1 organic lemon

1 clove garlic, crushed

2 tablespoons extra-virgin olive oil

½ teaspoon sea salt

1. Place the fish fillets in a shallow dish and pour the lemon juice over them. Marinate for at least 30 minutes or (preferably) overnight.

2. Preheat the oven to 400°F. Line a baking sheet with bleach-free parchment paper. Drain the lemon juice from the fish and pat the fillets dry with a paper towel. Place the fish on the baking sheet.

3. In a food processor, combine the pistachios, sage, cheese, lemon peel, garlic, olive oil, and salt. Process until smooth. Spread the pistachio mixture evenly over the fish fillets. Bake for 10 minutes, or until the fish flakes easily. Serve either immediately or at room temperature.

Easy Baked Miso Fish

(DF) (GF)   MAKES 2 SERVINGS

1 tablespoon brown rice vinegar   2 teaspoons honey

1 tablespoon organic white miso   2 fish fillets (4 ounces each) (see DNA

2 teaspoons tamari   Restart fish choices on page 102)

1. In a small bowl, combine the brown rice vinegar, miso, tamari, and honey. Whisk until well mixed. Pour half of the mixture into a shallow baking pan. Place the fish fillets in the pan, skin side down. Pour the remaining marinade over the fish and refrigerate overnight.

2. An hour before cooking, remove the fish from the refrigerator and allow to come to room temperature. Preheat the oven to 250°F. Line a baking dish with bleach-free parchment paper and place the fish fillets in the dish. Bake for 25 to 30 minutes, or until the fish flakes easily.

DNA Restart Food Tip: More than just a great source of minerals like manganese and zinc, which are needed to protect your DNA from oxidative stress, DNA Restart–approved fish are also a rich source of marine omega-3s and are lower in DNA-damaging contaminants like methylmercury. The miso in this recipe provides a fantastic umami hit, which will help you to feel satiated faster and for longer.

Saffron Chicken with
Vegetable Patch Stew and Almonds

(DF) (GF)   MAKES 4 SERVINGS (0.25 cc per serving)

2 medium chicken breasts, sliced into strips

½ cup Antioxidant Protein Marinade (page 238)

Pinch or 25 threads of saffron stigmas (not powder)

¼ cup boiling water

4 cloves garlic

1½ tablespoons extra-virgin olive oil, divided

1 medium onion, chopped

2 medium potatoes, chopped

2 tomatoes, chopped

2 red bell peppers, chopped

2 carrots, chopped

1 cup water

1 tablespoon double-concentrate tomato paste

Sea salt and freshly ground black pepper to taste

4 teaspoons almonds, with skins intact, roughly chopped

1. Place the chicken and marinade in a shallow bowl and cover with either a plate or plastic wrap. Marinate in the refrigerator for at least 2 hours or overnight.

2. Place the saffron stigmas in a small bowl or cup. Add the boiling water and allow to soak and infuse for 30 minutes. Meanwhile, crush the garlic and coarsely chop it. Set aside for 5 minutes.

3. In a stewing pan over medium heat, heat 1 tablespoon of water and 1 tablespoon of the olive oil. Remove the chicken from the marinade. When the pan is hot, sear the chicken on both sides. Add the garlic and onion and cook for 3 minutes. Reduce the heat to medium-low and add the potatoes, tomatoes, bell peppers, carrots, water, tomato paste, and saffron along with the infusion water. Cook for 1 hour, or until the vegetables are tender and the chicken is no longer pink. Season to taste with salt and black pepper. Drizzle with the remaining ½ tablespoon olive oil and garnish with the almonds before serving.

DNA Restart Food Tip: Saffron contains the carotenoid phytonutrients crocin and crocetin that have been shown to have antitumor and antioxidant effects. These compounds can even improve the symptoms and the effects of depression, premenstrual syndrome, and even excessive snacking behaviors.

Ceylon Cinnamon Beef Stew

(DF) (GF)   MAKES 8 SERVINGS

2 pounds pastured, organic stewing beef, cubed

½ cup Antioxidant Protein Marinade (page 238)

2 cups water

2 carrots, chopped

2 onions, chopped

5 cloves garlic, crushed and set aside for 5 minutes

1 medium tomato, chopped

1 teaspoon ground Ceylon cinnamon

Sea salt and freshly ground black pepper to taste

1 small bunch parsley (stems removed), finely chopped

1. In a medium bowl, cover the beef with the marinade and refrigerate overnight.

2. Drain and discard the marinade. Place the beef in a large stewing pot with the water and bring to a boil over medium heat. As scum rises to the surface, skim it off and discard until none remains. Simmer, covered, for 1 hour, adding more water if needed to keep the beef covered.

3. Add the carrots, onions, garlic, tomato, and cinnamon. Cook for 1 hour, covered, or until the beef is meltingly tender. Season to taste with salt and pepper. For maximum flavor, allow the stew to rest overnight in the refrigerator and reheat before serving. Garnish with the parsley before serving.

Greek Lamb Chops

(DF) (GF)   MAKES 4 SERVINGS

6 tablespoons fresh lemon juice

2 cloves garlic, crushed and set aside for 5 minutes

2 tablespoons dried Greek oregano

2 tablespoons extra-virgin olive oil

1 teaspoon sea salt

½ teaspoon freshly ground black pepper

4 grass-fed/pastured lamb loin chops (3 ounces each), trimmed

1. In a food processor, combine the lemon juice, garlic, oregano, olive oil, salt, and pepper. Process until smooth. Place the lamb chops in a shallow dish and pour the marinade over them. Marinate, covered, in the refrigerator overnight.

2. Preheat the oven to 350°F. Line a baking sheet with bleach-free parchment paper. Drain and discard the marinade from the chops. Place the chops on the baking sheet and bake for 20 minutes, turning once, or until browned and a thermometer inserted in the center registers 145°F for medium-rare.

ACKNOWLEDGMENTS

So many people played a significant role in supporting and bringing this project into reality, and I apologize in advance to any individual whom I may forget to mention below. A project of such scope and scale as what was required to bring *The DNA Restart* into fruition was assisted by an incredible number of people around the world over the last 2 decades.

To all of my scientific, research, and medical colleagues, who played such an important part in encouraging my scientific discovery and research over the years, and in so doing, supporting a lifetime pursuit of exploring the intersection of genetics, nutrition, and culture. It all started with *Apis mellifera ligustica*; they are, in the words of Gregor Mendel, "my dearest little animals."

To Woolloomooloo, more than just a protective harbor in the Western Pacific, you've managed to provide me with a practical base for all of my trips in Asia-Pacific over the years while doing research. You've all become a second family, and I want to particularly thank Jimmy, Mignon, Julie, Rachel, Jackie, and Michael for your professionalism, hard work, and efforts.

To Dr. Chiu Chui-Feng and the research staff of the Taiwan Tea Research and Extension Station, whom I met while visiting the center. Members of the Sri Lanka Tea Board were helpful for their insights into the history of Ceylon tea production and the history of *Camellia sinensis L.* cultivation on the island.

From the Centro Internacional de la Papa: Maria Elena Lanatta at CIP was instrumental in expertly organizing and planning my many visits with staff, as well as Drs. Gabriela Burgos, Oscar Oritz, Gordon Prain, Severin Polreich, Rainer Vollmer, and Thomas zum Felde, who were all incredibly illuminating in discussing their latest agricultural and genetics research projects in depth. All of your work regarding Andean root crops such as potatoes, sweet potatoes, and oxalis was fascinating to discuss. I was especially impressed by your efforts to protect Andean biodiversity and provide food security for us all by preserving and promoting native potato varieties through repatriation of disease-free cultivars. All of your significant efforts to alleviate poverty and work with local small-scale farmers in many parts of the world are inspiring to behold. Alejandro Argumedo was my very knowledgeable guide through the Sacred Valley and at the Peruvian Potato Park. It was truly exciting to see the work being done to protect and manage local genetic resources and traditional knowledge about food, health, and agriculture within the Sacred Valley and at the Peruvian Potato Park.

To the Guardians of the Potato, thank you all for your hospitality and generosity. Being your guest in the Altiplano and, of course, tasting your high-altitude potatoes was an incredible experience that I will not soon forget. As well to Asociación ANDES for all your coordinating and scheduling efforts.

To all of the talented and inspired chefs the world over, including:

Daniel Humm of the restaurant Eleven Madison Park, whose passion for delicious food and genuinely positive attitude about the importance of slowing down to enjoy exceptional food gave me a lot to think about.

David Kinch of Manresa, for his impressively precise articulation about his use of distinct flavors in the kitchen and his original thoughts on the ethical imperative of fusing healthy food with nutritional considerations.

Corey Lee of Benu, for his fascinating insights concerning the Asian vegetarian culinary tradition.

Pedro Miguel Schiaffino of Malabar Restaurant, who continues to inspire me with his zeal for novel flavors and ways of eating and living more healthfully.

Flavio Solorzano of El Señorío de Sulco for introducing me to the colorful world of Peruvian quinoa and of course sharing his ceviche secrets with me.

And to Yoshihiro Murata of Kikunoi, whose peerless knowledge of the *Kaiseki* culinary tradition puts him in a class of his own. All the chefs gave so generously of their time to share not only their creative gustatory genius, but the inner workings of their kitchens as well. A resounding thank-you to all.

A special thanks to Nobu Matsuhisa. I, too, will never forget that first evening we met at Nobu Tokyo—it was really such a blast! Thanks for sharing your early childhood memories and wide-ranging knowledge concerning the undeniable inner deliciousness of Japanese cuisine and, of course, your love of good food.

Many thanks to Dr. Kumiko Ninomiya of the Umami Information Center in Tokyo for graciously immersing me in the very substantive world of umami with all of its attending scientific complexities. Kumiko, Nobu was right. You truly are the "Umami Mama." And thanks to Dr. Ana San Gabriel, for further illuminating discussions on the dynamic genetic underpinnings of umami taste perception.

Ruth Klahsen of Monforte Dairy, a chef turned cheesemaker extraordinaire who generously shared her journey and delicious art. It was such a pleasure getting to know someone so passionate about and skilled in artisanal cheese production. Thomas Wilson from Spirit Tree Estate Cidery for sharing his encyclopedic knowledge regarding the history behind hard cider and the joys and true spirit behind his impressive production.

To Drs. Bruce Ames and Mark Tarnopolsky for generously giving of their time and scientific knowledge and speaking so candidly about their impressive research programs and careers. To Dr. Alexander Kokkinos for some enlivening conversations in Athens about the physiological and metabolic importance of Slow Living from not only a medical perspective but a cultural one as well.

And to my stellar publishing house, Rodale Books:

A special thank-you to Mary Ann Naples, for being such an overwhelming champion of The DNA Restart from the get-go. I knew from our first meeting at Rodale that you would have an instrumental role to play in launching this project to immeasurable heights. An exceptional thank-you to my very talented editor at Rodale, Jennifer Levesque, whose patience, suggestions, hard work, and probing questions were strongly influential in shaping the final book to be what it is today. Jennifer, you assembled the dream team to provide the necessary finishing touches to beautifully ready this book for press—thank you. And to Jean Lee for helping to keep us organized and on track. As well to the managing and copyediting and support team: Hope Clarke and Amy Kovalski, who with laser-sharp precision upgraded the manuscript immensely. Christina Gaugler, who did a stand-out job with the illustrations for the DNA Restart. To the marketing and publicity crew at Rodale that included Sindy Berner, Melissa Miceli, and Yelena Nesbit, who did a brilliant and creative job generating the deafening buzz around this book.

To the folks at 3Arts: Richard Abate, this is now our second book together, and it just keeps getting better. Thanks for keeping me laughing every step of the way. Also to Melissa Khan and Rachel Kim at 3Arts for keeping mine and Richard's lines of communication always open and perfectly synchronized.

To all of my dedicated and hardworking staff: my appreciation for the extra efforts you made to keep me up to date of our research and discovery progress while I was away traveling and doing research for The DNA Restart. To my executive assistant Claire Matthews, who somehow managed to keep my schedule organized across multiple time zones and dozens of cities over the last 2 years and not only ensured that I made all of my flights but also kept me seamlessly updated and informed while on the road.

To Sarah McDermmit, who served as the research coordinator and worked tirelessly to recruit our DNA Restarters and made sure to keep tabs on all of them as they transitioned through all 28 days and beyond. And, of course, last but not least, to all of the DNA Restarters themselves who provided continuous feedback along their weight-loss journeys, while also keeping immaculate notes and informing us of all of the varying transformations they were experiencing.

And to my parents, who instilled in me an early appreciation for the value of home cooking and always ensuring a beautiful weekly slow meal. To all of my family and friends who've been such a rich source of support over the years, even though my travels and research have kept my schedule tight, I cherish the time we are able to spend together. To my brother, Nunz, for his love, loyalty, and unwavering support. And to Shira, for everything.

BIBLIOGRAPHY/REFERENCES

The DNA Restart 1st Pillar: Eat for Your Genes

Borrell F, Junno A, Barceló JA. (2015). Synchronous environmental and cultural change in the emergence of agricultural economies 10,000 years ago in the Levant. *PLoS One* 10(8): e0134810.

Cani PD. (2015). Metabolism: dietary emulsifiers—sweepers of the gut lining? *Nature Reviews Endocrinology* 11(6): 319–20.

Chassaing B, Koren O, Goodrich JK, Poole AC, Srinivasan S, Ley RE, Gewirtz AT. (2015). Dietary emulsifiers impact the mouse gut microbiota promoting colitis and metabolic syndrome. *Nature* 519(7541): 92–6.

Elli L, Roncoroni L, Bardella MT. (2015). Non-celiac gluten sensitivity: time for sifting the grain. *World Journal of Gastroenterology* 21(27): 8221–6.

Falchi M, El-Sayed Moustafa JS, Takousis P, Pesce F, Bonnefond A, Andersson-Assarsson JC, Sudmant PH, et al. (2014). Low copy number of the salivary amylase gene predisposes to obesity. *Nature Genetics* 46(5): 492–7.

Feder JN, Gnirke A, Thomas W, Tsuchihashi Z, Ruddy DA, Basava A, Dormishian F, et al. (1996). A novel MHC class I-like gene is mutated in patients with hereditary haemochromatosis. *Nature Genetics* 13(4): 399–408.

Gerbault P, Roffet-Salque M, Evershed RP, Thomas MG. (2013). How long have adult humans been consuming milk? *International Union of Biochemistry and Molecular Biology* 65(12): 983–90.

Greenhill C. (2014). Obesity. Copy number variants in AMY1 connected with obesity via carbohydrate metabolism. *Nature Reviews Endocrinology* 10(6): 312.

Haag LM, Siegmund B. (2015). Intestinal microbiota and the innate immune system—a crosstalk in Crohn's disease pathogenesis. *Frontiers in Immunology* 6: 489.

Hoebler C, Karinthi A, Devaux MF, Guillon F, Gallant DJ, Bouchet B, Melegari C, Barry JL. (1998). Physical and chemical transformations of cereal food during oral digestion in human subjects. *British Journal of Nutrition* 80(5): 429–36.

Husby S, Murray J. (2015). Non-celiac gluten hypersensitivity: what is all the fuss about? *F1000 Prime Reports* 7: 54.

Imamura F, O'Connor L, Ye Z, Mursu J, Hayashino Y, Bhupathiraju SN, Forouhi NG. (2015). Consumption of sugar sweetened beverages, artificially sweetened beverages, and fruit juice and incidence of type 2 diabetes: systematic review, meta-analysis, and estimation of population attributable fraction. *British Medical Journal* 351: h3576.

Iron Disorders Institute: More information is available on hereditary hemochromatosis at the following Web site: http://www.hemochromatosis.org/.

Koeth RA, Wang Z, Levison BS, Buffa JA, Org E, Sheehy BT, Britt EB, et al. (2013). Intestinal microbiota metabolism of L-carnitine, a nutrient in red meat, promotes atherosclerosis. *Nature Medicine* 19(5): 576–85.

Kohlmeier M. (2013). *Nutrigenetics: applying the science of personal nutrition.* Oxford: Academic Press.

Krüttli A, Bouwman A, Akgül G, Della Casa P, Rühli F, Warinner C. (2014). Ancient DNA analysis reveals high frequency of European lactase persistence allele (T-13910) in medieval central Europe. *PLoS One* 9(1): e86251.

Leone VA, Cham CM, Chang EB. (2014). Diet, gut microbes, and genetics in immune function: can we leverage our current knowledge to achieve better outcomes in inflammatory bowel diseases? *Current Opinion in Immunology* 31: 16–23.

Losowsky MS. (2008). A history of coeliac disease. *Digestive Diseases* 26(2): 112–20.

Mandel AL, Breslin PA. (2012). High endogenous salivary amylase activity is associated with improved glycemic homeostasis following starch ingestion in adults. *Journal of Nutrition* 142(5): 853–8.

Marques-Bonet T, Kidd JM, Ventura M, Graves TA, Cheng Z, Hillier LW, Jiang Z, et al. (2009). A burst of segmental duplications in the genome of the African great ape ancestor. *Nature* 457: 877–81.

Mejía-Benítez MA, Bonnefond A, Yengo L, Huyvaert M, Dechaume A, Peralta-Romero J, Klünder-Klünder M, et al. (2015). Beneficial effect of a high number of copies of salivary amylase AMY1 gene on obesity risk in Mexican children. *Diabetologia* 58(2): 290–4.

Moalem S, Babul-Hirji R, Stavropolous DJ, Wherrett D, Bägli DJ, Thomas P, Chitayat D. (2012). XX male sex reversal with genital abnormalities associated with a de novo SOX3 gene duplication. *American Journal of Medical Genetics Part A* 158A(7): 1759–64.

Moalem S, Percy ME, Kruck TP, Gelbart RR. (2002). Epidemic pathogenic selection: an explanation for hereditary hemochromatosis? *Medical Hypotheses* 59(3): 325–9.

Moalem S, Prince JM. (2007). *Survival of the sickest: a medical maverick discovers why we need disease.* New York: William Morrow.

Moalem S, Weinberg ED, Percy ME. (2004). Hemochromatosis and the enigma of misplaced iron: implications for infectious disease and survival. *Biometals* 17(2): 135–9.

Naidoo N, Pawitan Y, Soong R, Cooper DN, Ku CS. (2011). Human genetics and genomics a decade after the release of the draft sequence of the human genome. *Human Genomics* 5(6): 577–622.

Perry GH, Dominy NJ, Claw KG, Lee AS, Fiegler H, Redon R, Werner J, et al. (2007). Diet and the evolution of human amylase gene copy number variation. *Nature Genetics* 39(10): 1256–60.

Salgia RJ, Brown K. (2015). Diagnosis and management of hereditary hemochromatosis. *Clinical Liver Disease* 19(1): 187–98.

Santos JL, Saus E, Smalley SV, Cataldo LR, Alberti G, Parada J, Gratacòs M, Estivill X. (2012). Copy number polymorphism of the salivary amylase gene: implications in human nutrition research. *Journal of Nutrigenetics and Nutrigenomics* 5(3): 117–31.

Sekirov I, Russell SL, Antunes LC, Finlay BB. (2010). Gut microbiota in health and disease. *Physiological Reviews* 90(3): 859–904.

Shiby VK, Mishra HN. (2013). Fermented milks and milk products as functional foods—a review. *Critical Reviews in Food Science and Nutrition* 53(5): 482–96.

Song M, Garrett WS, Chan AT. (2015). Nutrients, foods, and colorectal cancer prevention. *Gastroenterology* 148(6): 1244–60.

Tang WH, Hazen SL. (2014). The contributory role of gut microbiota in cardiovascular disease. *Journal of Clinical Investigation* 124(10): 4204–11.

Volta U, Caio G, De Giorgio R, Henriksen C, Skodje G, Lundin KE. (2015). Non-celiac gluten sensitivity: a work-in-progress entity in the spectrum of wheat-related disorders. *Best Practice & Research: Clinical Gastroenterology* 29(3): 477–91.

The DNA Restart 2nd Pillar: Reverse Aging

Afshin A, Micha R, Khatibzadeh S, Mozaffarian D. (2014). Consumption of nuts and legumes and risk of incident ischemic heart disease, stroke, and diabetes: a systematic review and meta-analysis. *American Journal of Clinical Nutrition* 100: 278–88.

Alasalvar C, Bolling BW. (2015). Review of nut phytochemicals, fat-soluble bioactives, antioxidant components and health effects. *British Journal of Nutrition* 113 Suppl 2: S68–78.

Arreola R, Quintero-Fabián S, López-Roa RI, Flores-Gutiérrez EO, Reyes-Grajeda JP, Carrera-Quintanar L, Ortuño-Sahagún D. (2015). Immunomodulation and anti-inflammatory effects of garlic compounds. *Journal of Immunology Research* 401630. Epub.

Bao W, Rong Y, Rong S, Liu L. (2012). Dietary iron intake, body iron stores, and the risk of type 2 diabetes: a systematic review and meta-analysis. *BMC Medicine* 10: 119.

Bazzano LA, He J, Ogden LG, Loria C, Vupputuri S, Myers L, Whelton PK. (2001). Legume consumption and risk of coronary heart disease in US men and women: NHANES I Epidemiologic Follow-up Study. *Archives of Internal Medicine* 161(21): 2573–78.

Bazzano LA, Thompson AM, Tees MT, Nguyen CH, Winham DM. (2011). Non-soy legume consumption lowers cholesterol levels: a meta-analysis of randomized controlled trials. *Nutrition, Metabolism and Cardiovascular Diseases* 21(2): 94–103.

Bouchenak M, Lamri-Senhadji M. (2013). Nutritional quality of legumes, and their role in cardiometabolic risk prevention: a review. *Journal of Medicinal Food* 16(3): 185–98.

Bown S. (2003). *Scurvy: how a surgeon, a mariner and a gentleman solved the greatest medical mystery of the age of sail.* West Sussex: Summersdale Publishing Ltd.

Chen S, Shen X, Cheng S, Li P, Du J, Chang Y, Meng H. (2013). Evaluation of garlic cultivars for polyphenolic content and antioxidant properties. *PLoS One* 8(11): e79730.

Crane JD, MacNeil LG, Lally JS, Ford RJ, Bujak AL, Brar IK, Kemp BE, Raha S, Steinberg GR, Tarnopolsky MA. (2015). Exercise-stimulated interleukin-15 is controlled by AMPK and regulates skin metabolism and aging. *Aging Cell* 14(4): 625–34.

Devries MC, Samjoo IA, Hamadeh MJ, McCready C, Raha S, Watt MJ, Steinberg GR, Tarnopolsky MA. (2013). Endurance training modulates intramyocellular lipid compartmentalization and morphology in skeletal muscle of lean and obese women. *Journal of Clinical Endocrinology & Metabolism* 98(12): 4852–62.

EFSA (2011). The European Food Safety Authority. Scientific opinion on the substantiation of health claims related to polyphenols in olive and protection of LDL particles from oxidative damage (ID 1333, 1638, 1639, 1696, 2865), maintenance of normal blood HDL cholesterol concentrations (ID 1639), maintenance of normal blood pressure (ID 3781), "anti-inflammatory properties" (ID 1882), "contributes to the upper respiratory tract health" (ID 3468), "can help to maintain a normal function of gastrointestinal tract" (3779), and "contributes to body defences against external agents" (ID 3467) pursuant to article 13(1) of regulation (ec) no 1924/2006. *EFSA Journal* 9: 2033–57.

Franceschi C, Campisi J. (2014). Chronic inflammation (inflammaging) and its potential contribution to age-associated diseases. *Journals of Gerontology Series A: Biological Sciences and Medical Sciences* 69 Suppl 1:S4–9.

Frasca D, Blomberg BB. (2015). Inflammaging decreases adaptive and innate immune responses in mice and humans. *Biogerontology* [Epub ahead of print].

Gibis M, Weiss J. (2012). Antioxidant capacity and inhibitory effect of grape seed and rosemary extract in marinades on the formation of heterocyclic amines in fried beef patties. *Food Chemistry* 134: 766–74.

Gillen JB, Percival ME, Skelly LE, Martin BJ, Tan RB, Tarnopolsky MA, Gibala MJ. (2014). Three minutes of all-out intermittent exercise per week increases skeletal muscle oxidative capacity and improves cardiometabolic health. *PLoS One* 9(11): e111489.

Glaser N, Stopper H. (2012). Patulin: mechanism of genotoxicity. *Food and Chemical Toxicology* 50(5): 1796–801.

Gupta H. (2014). Barriers to and facilitators of long term weight loss maintenance in adult UK people: a thematic analysis. *International Journal of Preventive Medicine* 5(12): 1512–20.

Ha V, Sievenpiper JL, de Souza RJ, Jayalath VH, Mirrahimi A, Agarwal A, Chiavaroli L, et al. (2014). Effect of dietary pulse intake on established therapeutic lipid targets for cardiovascular risk reduction: a systematic review and meta-analysis of randomized controlled trials. *Canadian Medical Association Journal* 186(8): E252–62.

Han YM, Park JM, Jeong M, Yoo JH, Kim WH, Shin SP, Ko WJ, Hahm KB. (2015). Dietary, non-microbial intervention to prevent Helicobacter pylori-associated gastric diseases. *Annals of Translational Medicine* 3(9): 122.

Hermsdorff HH, Zulet MÁ, Abete I, Martínez JA. (2011). A legume-based hypocaloric diet reduces proinflammatory status and improves metabolic features in overweight/obese subjects. *European Journal of Nutrition* 50(1): 61–9.

Hwang KO, Etchegaray JM, Sciamanna CN, Bernstam EV, Thomas EJ. (2014). Structural social support predicts functional social support in an online weight loss programme. *Health Expectations* 17(3): 345–52.

Kohlmeier M. (2013). *Nutrigenetics: applying the science of personal nutrition.* Oxford: Academic Press.

Lavie CJ, Lee DC, Sui X, Arena R, O'Keefe JH, Church TS, Milani RV, Blair SN. (2015). Effects of running on chronic diseases and cardiovascular and all-cause mortality. *Mayo Clinic Proceedings* pii: S0025-6196(15): 621–27.

Lichtfouse E (ed.). (2012). *Organic fertilisation, soil quality and human health.* UK: Springer.

Mattson MP. (2015). Lifelong brain health is a lifelong challenge: from evolutionary principles to empirical evidence. *Aging Research Reviews* 20: 37–45.

Messina V. (2014). Nutritional and health benefits of dried beans. *American Journal of Clinical Nutrition* 100 Suppl 1: 437S–42S.

Meucci V, Razzuoli E, Soldani G, Massart F. (2010). Mycotoxin detection in infant formula milks in Italy. *Food Additives & Contaminants. Part A, Chemistry, Analysis, Control, Exposure & Risk Assessment* 27(1): 64–71.

Milone M, Massie R. (2010). Polymerase gamma 1 mutations: clinical correlations. *Neurologist* 16(2): 84–91.

Moalem S, LaPlante ME. (2014). *Inheritance: how our genes change our lives and our lives change our genes.* New York: Grand Central Publishing.

Mumme K, Stonehouse W. (2015). Effects of medium-chain triglycerides on weight loss and body composition: a meta-analysis of randomized controlled trials. *Journal of the Academy of Nutrition and Dietetics* 115(2): 249–63.

O'Donovan A, Pantell MS, Puterman E, Dhabhar FS, Blackburn EH, Yaffe K, Cawthon RM, et al. (2011). Cumulative inflammatory load is associated with short leukocyte telomere length in the Health, Aging and Body Composition Study. *PLoS One* 6: e19687.

Poma A, Fontecchio G, Carlucci G, Chichiriccò G. (2012). Anti-inflammatory properties of drugs from saffron crocus. *Anti-inflammatory & Anti-allergy Agents in Medicinal Chemistry* 11(1): 37–51.

Poulsen MW, Hedegaard RV, Andersen JM, de Courten B, Bugel S, Nielsen J, Skibsted LH, Dragsted LO. (2013). Advanced glycation endproducts in food and their effects on health. *Food and Chemical Toxicology* 60: 10–37.

Raiola A, Tenore GC, Manyes L, Meca G, Ritieni A. (2015). Risk analysis of main mycotoxins occurring in food for children: an overview. *Food and Chemical Toxicology* 84: 169–80.

Rochfort S, Panozzo J. (2007). Phytochemicals for health, the role of pulses. *Journal of Agriculture and Food Chemistry* 55: 7981–94.

Safdar A, Bourgeois JM, Ogborn DI, Little JP, Hettinga BP, Akhtar M, Thompson JE, et al. (2011). Endurance exercise rescues progeroid aging and induces systemic mitochondrial rejuvenation in mtDNA mutator mice. *Proceedings of the National Academy of Sciences* 108(10): 4135–40.

Samitz G, Egger M, Zwahlen M. (2011). Domains of physical activity and all-cause mortality: systematic review and dose-response meta-analysis of cohort studies. *International Journal of Epidemiology* 40(5): 1382–1400.

Samjoo IA, Safdar A, Hamadeh MJ, Raha S, Tarnopolsky MA. (2013). The effect of endurance exercise on both skeletal muscle and systemic oxidative stress in previously sedentary obese men. *Nutrition & Diabetes* 3: e88.

Sanlorenzo M, Wehner MR, Linos E, Kornak J, Kainz W, Posch C, Vujic I, et al. (2015). The risk of melanoma in airline pilots and cabin crew: a meta-analysis. *JAMA Dermatology* 151(1): 51–58.

Stubbs B, Rosenbaum S, Vancampfort D, Ward PB, Schuch FB. (2015). Exercise improves cardiorespiratory fitness in people with depression: a meta-analysis of randomized control trials. *Journal of Affective Disorders* 190: 249–53.

Toussaint-Samat M. (1993). *History of food*. Cambridge: Blackwell Publishers.

Turati F, Pelucchi C, Guercio V, La Vecchia C, Galeone C. (2015). Allium vegetable intake and gastric cancer: a case-control study and meta-analysis. *Molecular Nutrition & Food Research* 59(1): 171–79.

Winham DM, Hutchins AM, Johnston CS. (2007). Pinto bean consumption reduces biomarkers for heart disease risk. *Journal of the American College of Nutrition* 26(3): 243–9.

Winters-Stone K. (2015). Exercise and cancer risk—how much is enough? *JAMA Oncology* 1(6): 776–7.

Zeisel SH. (2006). Choline: critical role during fetal development and dietary requirements in adults. *Annual Review of Nutrition* 26: 229–50.

Zeisel SH. (2012). Dietary choline deficiency causes DNA strand breaks and alters epigenetic marks on DNA and histones. *Mutation Research* 733(1–2): 34–38.

Zeisel SH, da Costa KA. (2009). Choline: an essential nutrient for public health. *Nutrition Review* 67(11): 615–23.

The DNA Restart 3rd Pillar: Eat Umami

Anthony M, Blumenthal H, Bourdas A, Kinch D, Martinez V, Matsuhisa N, Murata Y, et al. (2014). *Umami: the fifth taste*. Tokyo: Japan Publications Trading Co.

Bauer LB, Reynolds LJ, Douglas SM, Kearney ML, Hoertel HA, Shafer RS, Thyfault JP, Leidy HJ. (2015). A pilot study examining the effects of consuming a high-protein vs normal-protein breakfast on free-living glycemic control in overweight/obese 'breakfast skipping' adolescents. *International Journal of Obesity* 39(9): 1421–4.

Cardel M, Lemas DJ, Jackson KH, Friedman JE, Fernández JR. (2015). Higher intake of PUFAs is associated with lower total and visceral adiposity and higher lean mass in a racially diverse sample of children. *Journal of Nutrition* 145(9): 2146–52.

Chaudhari N, Roper SD. (2010). The cell biology of taste. *Journal of Cell Biology* 190(3): 285–96.

Fares H, Lavie CJ, DiNicolantonio JJ, O'Keefe JH, Milani RV. (2014). Omega-3 fatty acids: a growing ocean of choices. *Current Atherosclerosis Reports* 16(2): 389.

Igarashi M, Santos RA, Cohen-Cory S. (2015). Impact of maternal n-3 polyunsaturated fatty acid deficiency on dendritic arbor morphology and connectivity of developing Xenopus laevis central neurons in vivo. *Journal of Neuroscience* 35(15): 6079–92.

Kurihara K. (2015). Umami the fifth basic taste: history of studies on receptor mechanisms and role as a food flavor. *Journal of Physiology* [Epub ahead of print].

Montgomery P, Burton JR, Sewell RP, Spreckelsen TF, Richardson AJ. (2014). Fatty acids and sleep in UK children: subjective and pilot objective sleep results from the DOLAB study—a randomized controlled trial. *Journal of Sleep Research* 23(4): 364–88.

Mouritsen OG, Styrbæk K. (2014). *Umami: unlocking the secrets of the fifth taste*. New York: Columbia University Press.

Oruna-Concha MJ, Methven L, Blumenthal H, Young C, Mottram DS. (2007). Differences in glutamic acid and 5'-ribonucleotide contents between flesh and pulp of tomatoes and the relationship with umami taste. *Journal of Agriculture and Food Chemistry* 55(14): 5776–80.

Quote from Heather Leidy, PhD, from the University of Missouri School of Medicine. The interview quote is archived in the following site: http://www.sciencedaily.com/releases/2015/08/150812165923.htm.

Roper SD, Chaudhari N. (2009). Processing umami and other tastes in mammalian taste buds. *Annals of the New York Academy of Sciences* 1170: 60–5.

Rotola-Pukkila MK, Pihlajaviita ST, Kaimainen MT, Hopia AI. (2015). Concentration of umami compounds in pork meat and cooking juice with different cooking times and temperatures. *Journal of Food Science* [Epub ahead of print].

Sasano T, Satoh-Kuriwada S, Shoji N, Iikubo M, Kawai M, Uneyama H, Sakamoto M. (2014). Important role of umami taste sensitivity in oral and overall health. *Current Pharmaceutical Design* 20(16): 2750–4.

Toussaint-Samat M. (1993). *History of food*. Cambridge: Blackwell Publishers.

The DNA Restart 4th Pillar: Drink Oolong Tea

Bajzer M, Steeley RJ. (2006). Physiology: obesity and gut flora. *Nature* 444: 1009–10.

Chen J, Qin S, Xiao J, Tanigawa S, Uto T, Hashimoto F, Fujii M, Hou DX. (2011). A genome-wide microarray highlights the antiinflammatory genes targeted by oolong tea theasinensin A in macrophages. *Nutrition and Cancer* 63(7): 1064–73.

Chitpan M, Wang X, Ho CT, Huang Q. (2007). Monitoring the binding processes of black tea thearubigin to the bovine serum albumin surface using quartz crystal microbalance with dissipation monitoring. *Journal of Agriculture and Food Chemistry* 55(25): 10110–6.

De Filippo C, Cavalieri D, Di Paola M, Ramazzotti M, Poullet JB, Massart S, Collini S, Pieraccini G, Lionetti P. (2010). Impact of diet in shaping gut microbiota revealed by a comparative study in children from Europe and rural Africa. *Proceedings of the National Academy of Sciences* 107(33): 14691–6.

Fei N, Zhao L. (2013). An opportunistic pathogen isolated from the gut of an obese human causes obesity in germfree mice. *The ISME Journal* 7: 880–4.

Forbes A, Henley D. (2011). Shennong in: *China's ancient tea horse road*. Chiang Mai: Cognoscenti Books.

Gautier L. (2006). Le thé, arômes & saveurs du monde. France: Aubanel.

Gorjanović S, Komes D, Pastor FT, Belščak-Cvitanović A, Pezo L, Hečimović I, Sužnjević D. (2012). Antioxidant capacity of teas and herbal infusions: polarographic assessment. *Journal of Agriculture and Food Chemistry* 60(38): 9573–80.

He RR, Chen L, Lin BH, Matsui Y, Yao XS, Kurihara H. (2009). Beneficial effects of oolong tea consumption on diet-induced overweight and obese subjects. *Chinese Journal of Integrative Medicine* 15(1): 34–41.

Heber D, Zhang Y, Yang J, Ma JE, Henning SM, Li Z. (2014). Green tea, black tea, and oolong tea polyphenols reduce visceral fat and inflammation in mice fed high-fat, high-sucrose obesogenic diets. *Journal of Nutrition* 144(9): 1385–93.

Heiss ML, Heiss RJ. (2010). *The tea enthusiast's handbook: a guide to the world's best teas*. Berkeley: Ten Speed Press.

Hisanaga A, Ishida H, Sakao K, Sogo T, Kumamoto T, Hashimoto F, Hou DX. (2014). Anti-inflammatory activity and molecular mechanism of Oolong tea theasinensin. *Food & Function* 5: 1891–97.

Hou DX, Masuzaki S, Tanigawa S, Hashimoto F, Chen J, Sogo T, Fujii M. (2010). Oolong tea theasinensins attenuate cyclooxygenase-2 expression in lipopolysaccharide (LPS)-activated mouse macrophages: structure-activity relationship and molecular mechanisms. *Journal of Agriculture and Food Chemistry* 58: 12735–43.

Hullar MAJ, Fu BC. (2014). Diet, the gut microbiome, and epigenetics. *Cancer Journal* 20(3): 170–5.

Hursel R, Westerterp-Plantenga MS. (2013). Catechin—and caffeine—rich teas for control of body weight in humans. *American Journal of Clinical Nutrition* 98: 1682S–93S.

Lee SJ, Jia Y. (2015). The effect of bioactive compounds in tea on lipid metabolism and obesity through regulation of peroxisome proliferator-activated receptors. *Current Opinion in Lipidology* 26(1): 3–9.

Legeay S, Rodier M, Fillon L, Faure S, Clere N. (2015). Epigallocatechin gallate: a review of its beneficial properties to prevent metabolic syndrome. *Nutrients* 7(7): 5443–68.

Lowe ME. (1997). Structure and function of pancreatic lipase and colipase. *Annual Review of Nutrition* 17: 141–58.

Moalem S, LaPlante ME. (2014). *Inheritance: how our genes change our lives and our lives change our genes*. New York: Grand Central Publishing.

Moalem S, Prince JM. (2007). *Survival of the sickest: a medical maverick discovers why we need disease*. New York: William Morrow.

O'Connor L, Imamura F, Lentjes MA, Khaw KT, Wareham NJ, Forouhi NG. (2015). Prospective associations and population impact of sweet beverage intake and type 2 diabetes, and effects of substitutions with alternative beverages. *Diabetologia* 58(7): 1474–83.

Rastmanesh R. (2011). High polyphenol, low probiotic diet for weight loss because of intestinal microbiota interaction. *Chemico-Biological Interactions* 189: 1–8.

Sekirov I, Russell SL, Antunes LC, Finlay BB. (2010). Gut microbiota in health and disease. *Physical Review Letters* 90: 859–904.

Seo DB, Jeong HW, Cho D, Lee BJ, Lee JH, Choi JY, Bae IH, Lee SJ. (2015). Fermented green tea extract alleviates obesity and related complications and alters gut microbiota composition in diet-induced obese mice. *Journal of Medicinal Food* 18(5): 549–56.

Taiwanese Research Center. (2014–5). Personal communication.

UNESCO. (2011). It took the author 27 years to compile the information included in the *Ben cao gang mu compendium of materia medica*. The following Web site (http://www.unesco.org) provides some more background information behind this fascinating text.

Villano D, Lettieri-Barbato D, Guadagni F, Schmid M, Serafini M. (2012). Effect of acute consumption of oolong tea on antioxidant parameters in healthy individuals. *Food Chemistry* 132: 2102–6.

Yang CS, Chen G, Qing Wu. (2014). Recent scientific studies of a traditional Chinese medicine, tea, on prevention of chronic diseases. *Journal of Traditional and Complementary Medicine* 4(1): 17–23.

The DNA Restart 5th Pillar: Slow Living

Blackburn EH. (2000). Telomere states and cell fates. *Nature* 408: 53–56.

Chaput JP, Després JP, Bouchard C, Tremblay A. (2008). The association between sleep duration and weight gain in adults: a 6-year prospective study from the Quebec Family Study. *Sleep* 31(4): 517–23.

Chaput JP, Tremblay A. (2007). Does short sleep duration favor abdominal adiposity in children? *International Journal of Pediatrics* 2(3): 188–91.

Dalen J, Smith BW, Shelley BM, Sloan AL, Leahigh L, Begay D. (2010). Pilot study: Mindful Eating and Living (MEAL): weight, eating behavior, and psychological outcomes associated with a mindfulness-based intervention for people with obesity. *Journal of Traditional and Complementary Medicine* 18(6): 260–4.

Dwyer L, Oh A, Patrick H, Hennessy E. (2015). Promoting family meals: a review of existing interventions and opportunities for future research. *Journal of Adolescent Health, Medicine and Therapeutics* 6: 115–31.

Epel ES, Blackburn EH, Lin J, Dhabhar FS, Adler NE, Morrow JD, Cawthon RM. (2004). Accelerated telomere shortening in response to life stress. *Proceedings of the National Academy of Sciences* (49): 17312–5.

Jackowska M, Hamer M, Carvalho LA, Erusalimsky JD, Butcher L, Steptoe A. (2012). Short sleep duration is associated with shorter telomere length in healthy men: findings from the Whitehall II cohort study. *PLoS One* 7(10): e47292.

Masri S, Kinouchi K, Sassone-Corsi P. (2015). Circadian clocks, epigenetics, and cancer. *Current Opinion in Oncology* 27(1): 50–56.

Ohkuma T, Hirakawa Y, Nakamura U, Kiyohara Y, Kitazono T, Ninomiya T. (2015). Association between eating rate and obesity: a systematic review and meta-analysis. *International Journal of Obesity* [Epub ahead of print—May 25, 2015].

O'Reilly GA, Cook L, Spruijt-Metz D, Black DS. (2015). Mindfulness-based interventions for obesity-related eating behaviours: a literature review. *Obesity Reviews* 15(6): 453–61.

Prather AA, Gurfein B, Moran P, Daubenmier J, Acree M, Bacchetti P, Sinclair E, et al. (2015). Tired telomeres: poor global sleep quality, perceived stress, and telomere length in immune cell subsets in obese men and women. *Brain, Behavior, and Immunity* 47: 155–62.

Qureshi IA, Mehler MF. (2014). Epigenetics of sleep and chronobiology. *Current Neurology and Neuroscience Reports* 14(3): 432–40.

Robinson E, Almiron-Roig E, Rutters F, de Graaf C, Forde CG, Tudur Smith C, Nolan SJ, Jebb SA. (2014). A systematic review and meta-analysis examining the effect of eating rate on energy intake and hunger. *American Journal of Clinical Nutrition* 100(1): 123–51.

Valdes AM, Andrew T, Gardner JP, Kimura M, Oelsner E, Cherkas LF, Aviv A, Spector TD. (2005). Obesity, cigarette smoking, and telomere length in women. *Lancet* 366(9486): 662–4.

Wolkowitz OM, Mellon SH, Epel ES, Lin J, Dhabhar FS, Su Y, Reus VI, et al. (2011). Leukocyte telomere length in major depression: correlations with chronicity, inflammation and oxidative stress—preliminary findings. *PLoS One* 6(3): e17837.

INDEX

B

Basal metabolism, oolong tea
    increasing, 175, 176
Belly fat. *See* Abdominal fat
*Ben Cao Gang Mu (Compendium of
    Materia Medica)*, 168
Biological home economics, 9–10
Bitter taste, genetic role of, 131
Black tea, 165, 166, 167, 172
Blane, Gilbert, 93
Blood tests, for iron assessment, 43,
    43
Brassica plants, raw, avoiding, 85
Breads
    carb cost of, 205
    emulsifiers and, 29–30
Breakfast
    skipping, 134–35
    umami for, 134, 135, 221
Breast cancer, causes of, 45, 51
Breast milk, umami in, 136
British sailors, vitamin C benefiting,
    92, 93
Brown fat, oolong tea activating, 176
Bubonic plague, 24

C

Caffeine
    in oolong tea, 174, 175, 176
    sensitivity to, 174, 180
Calcium, 88
Cancer
    alcohol and, 47, 48, 49, 51
    causes of
        choline deficiency, 87
        chronic inflammation, 197
        diet, 82
        DNA damage, 56, 69, 70, 74
        natural chemical poisons, 83
        peanuts, 110
        red meat, 45, 45, 179
        UV radiation, 70, 79, 80

preventing, with
    exercise, 59, 60, 61
    garlic and onion, 117–18
    omega-3 PUFAs, 97
    selenium, 89
Canker sores, 27–28, 30
Carb Cost Allowance Guide, 18,
    203–6
Carbohydrate breakdown, *AMY1* gene
    influencing, 8–9
Carbohydrate consumption
    AMY1 gene and, 10, 11
    case study on, xiii–xiv
    genetic influence on, 3, 6
    ideal
        finding, xiv, xv, 14, 16, 17, **17**
        meal plans for, 207–18
        percentage allowed, 16, 17, **17**
Carbohydrate Consumption
    Categories. *See* Full Carb
    Category; Moderate Carb
    Category; Restricted Carb
    Category
Carbohydrate Consumption Estimate
    Guide, 16, 17, **17**
Carbohydrates
    energy density of, 96
    gluten-free, 21
Cardiovascular disease. *See also* Heart
    disease
    preventing, 97, 105, 111, 228
    from red meat, 45–46, 179
Carnitine, 45
Carotenoids, 77, 106, 114, 119, 122,
    123
Carrageenans, dairy intolerance from,
    39–40
Case studies, 7
    Chad—coffee sensitivity, 163
    Claire—carbohydrate consumption,
        xiii–xiv
    Fiona—celiac disease and infertility,
        19–20, 21